National Defense Research Institute

CFE AND MILITARY STABILITY IN EUROPE

JOHN E. PETERS

Prepared for the
Office of the Secretary of Defense

RAND

PREFACE

RAND's research effort to provide analytic support over the past two years to the Office of Non-Nuclear Arms Control, Office of the Under Secretary of Defense for Policy, has ranged widely. First, it aided preparation for the CFE (Conventional Forces in Europe) Implementation Review Conference held in May 1996 and, more recently, reinforced U.S. negotiations in the CFE Adaptation Talks. Over the ensuing months, the project has explored U.S. negotiating options and the consequences associated with potential new foreign arms control proposals. This report is a record of our analytic support.

This project was conducted within the Center for International Security and Defense Policy of RAND's National Defense Research Institute (NDRI). NDRI is a federally funded research and development center sponsored by the Office of the Secretary of Defense, the Joint Staff, and the defense agencies.

This report should be of interest to specialists in European security or arms control.

CONTENTS

FIGURES

TABLES

SUMMARY

It is difficult, in 1997, to conceive of European conventional arms control as a major issue. After all, the Cold War is over. But it remains important for the 30 signatories of the Conventional Forces in Europe (CFE) Treaty and the former warring parties in Bosnia. For the CFE parties, the treaty has come to be seen as the baseline for European stability and a bellwether for the health of European international relations. For the Balkan contestants, conventional arms control contributed directly to whatever stability they enjoy under the Dayton Agreement. Seven years after the conclusion of the Cold War, Europe remains engaged in conventional arms control.

THE ORIGINAL OBJECTIVES OF CFE

CFE was well designed for the bipolar confrontation of the Cold War. The treaty controlled the weapons thought to be most militarily significant for offensive operations in any NATO–Warsaw Pact war. It caused both blocs of states to reduce their treaty-limited equipment (TLE) to equal amounts: 20,000 tanks, 20,000 artillery pieces, 30,000 armored combat vehicles, 6,800 combat aircraft, and 2,000 attack helicopters. It caused the participants to destroy some 50,000 pieces of TLE and to carry out some 2,700 inspections. The subzone structure of the treaty limited each bloc's ability to concentrate its TLE and produce the overwhelming force ratios desirable for offensive operations. Stability resulted, since neither side could generate a surprise attack while observing the limitations of the treaty.

To ensure that signatories complied with the treaty's provisions, intrusive verification and monitoring provisions were included.

Moreover, a series of confidence- and security-building measures (CSBMs) codified in the Vienna Document of 1994 reinforced the stability provisions of CFE by requiring states to invite observers to exercises above a certain size, to notify the other signatories 42 days in advance of exercises involving a certain number of troops, and to limit exercises involving 40,000 troops or 900 battle tanks to once every two years. These provisions dispelled suspicion about states' motives and activities and contributed to the prevention of surprise attack.

THE CURRENT ROLE OF CFE

Given the transformed security landscape in Europe, CFE in its current configuration is much less central to Europe's problems. The treaty does little to control strife between two European states: The subzones that were intended to prevent the massing of NATO and Warsaw Pact forces along the Inter-German Boundary do not prevent smaller concentrations in central Europe between neighbors with unsettled grievances. The flank provisions are a source of irritation between Russia and the West, since Moscow insists that its internal security requirements demand more TLE in the flank regions than the treaty allows. The other flank states, Turkey and Norway, have been adamant about Russian compliance.

The Vienna Document CSBMs requiring advance notice for exercises and observers under some circumstances still have important utility, but they, too, were designed to prevent the surprise massing of very large forces, not the smaller ones that threaten ethnic or interstate conflict in today's Europe. Similarly, the CFE inspection quotas and inspection system seem less adequate today because more states are interested in inspecting Russia and their immediate neighbors than they were before. In short, the inspection regime and counter-surprise features of the current CFE Treaty and Vienna Document do not adequately address current concerns.

Although CFE managed a major concern during the Cold War—the fear of surprise conventional attack—the treaty clearly plays a more limited role given the post–Cold War security agenda. At best, many of its measures have served as models for later security solutions. For example, many of the Dayton Accords' arms control features reflect the influence of CFE: zones of separation similar to CFE subzones, treaty-limited equipment, monitoring. In any case, the direct contribution to European security that CFE currently makes is much smaller than its contribution would have been during the Cold War era.

THE PROJECT'S RESEARCH AGENDA

For the past two years, this project has been helping the Secretary of Defense's Office of Non-Nuclear Arms Control grapple with the changes described above, first to prepare for the CFE Review Conference in May 1996 and subsequently to support CFE Adaptation Talks, currently under way in Vienna. Along the way, the project has examined legal-procedural issues that might determine the scope of change that the treaty signatories would be willing to embrace. The project has considered traditional military balances and has explored new ways to employ military metrics to evaluate and, perhaps, control current forces and the military infrastructure of participating states. The project has attempted exploratory modeling as an aid to thinking about and measuring the CFE Treaty's shortcomings in the present security environment. Finally, the project has considered the future of more-localized arms control as a potential next step for CFE. This report details the highlights of that research.

THE OUTLOOK

At this point, there is every reason to believe, unless the talks derail because of the NATO-Russia dispute over alliance enlargement, that the CFE Adaptation Talks will come to a positive conclusion. Two years of study suggest that, on the one hand, there is substantial flexibility in the United States' arms control repertoire and that, on the other hand, the "fit" between what CFE can deliver and Europe's current security requirements is loose enough so that it should be possible to preserve the essential elements of CFE as a "security redoubt" against the worst-case fall in Europe's fortunes while adjusting some aspects of the treaty to fit present circumstances more closely.

With these points in mind and based on the project's research results, we counsel in this report against undertaking additional pan-European conventional arms control initiatives. To the extent that arms control will be useful in the near future, it will involve more-local agreements tailored specifically to address grievances among neighbors. Unless circumstances alter dramatically, Europe-wide negotiations will make little sense, especially in the face of NATO enlargement, for which, presumably, allies will not negotiate arms control pacts with each other.

If the United States were pressed for more arms control (as it may be, since both the NATO-Russian Founding Act and the Ukraine-NATO Charter include specific expectations about how CFE will improve the parties' security in the wake of NATO expansion), this study indicates that stability and transparency measures could make important contributions. Deep TLE reductions, coupled with robust confidence- and security-building measures, could also have a major effect, although the expense of the reductions and the current political climate in much of Europe make deep reductions an unlikely choice.

What next? Assuming that CFE modernization is successful, what should the United States consider its next arms control priority? Some unfinished business remains on the European arms control agenda: bringing the Baltic states and several other states into CFE. And there is potential for arms control to help manage the dangers emerging from the Mediterranean basin and Southwest Asia. Membership in both the treaty and the CSBM regime could help insulate the Baltic states from Russian pressures and could provide transparency sufficient to defuse misinformation campaigns and rumors about poor treatment of Russian minorities and similar issues that might offer Moscow excuses to intervene. In addition, membership in the two pacts would entitle the Baltic states to information about their neighbors, including prior notice of major exercises and similar activities that, given the small size of these states, could be very dangerous to their prospects of survival as independent, sovereign actors.

The history of arms control in the Cold War offers some clues as to how arms control might be used beyond Europe's borders to improve relations with her neighbors. Arms control offered a forum for discussions with the Russians in an era when the West was ideologically, economically, and culturally alienated from the Soviet Union. In practice, arms control provided a sound channel of communication when few others were available to sustain a useful dialogue.

Through this channel, the East and West came gradually to know each other better. The United States faces similar ideological barriers in and around the Mediterranean basin. Radical Islam, cults of the leader, and similar forces promote barriers to peaceful relations analogous to those encountered with the Soviet Union. It therefore makes sense to use a proven tool—arms control—to try to manage the relationship with these actors and, over time, to reduce their radicalism.

THE NEAR-TERM CONVENTIONAL ARMS CONTROL AGENDA

Even if the establishment of an arms control dialogue like that suggested in this report proves bureaucratically impracticable, the arms control agenda remains full. For most of Europe and the United States, the agenda could be summarized as follows:

- Continue to participate in the usual arms control processes, consolidating CFE and the CSBM regime by incorporating the Baltic states.
- That accomplished, move the discussion from arms control to genuine cooperation.
- Consider mutually beneficial projects, perhaps a regional missile defense system, cooperative anti-submarine warfare patrolling in the Mediterranean, or accelerated work on NATO's Combined Joint Task Force concept.

The agenda for trouble spots in Europe is as follows:

- Where appropriate, the United States should promote solid, local arms control deals.
- As with the Dayton Accords, the United States should use its influence to bring disputing parties together for the negotiations.

ACKNOWLEDGMENTS

Colleague William O'Malley made valuable contributions over the course of the project. His insights were particularly helpful in crafting stability measures based on national infrastructure. Our principal contact in the sponsor's office, Mr. Steven Schleien, was an invaluable source of information and asker of tough questions. Mr. Dorn Crawford of the United States Arms Control and Disarmament Agency and COL Jeffrey D. McCausland of the U.S. Army War College provided insightful reviews of the manuscript. Any errors are my own.

ACRONYMS

AC	Aircraft
ACDA	Arms Control and Disarmament Agency
ACE	Allied Command, Europe
AD-AT	Air defense—anti-tank
APC-AIFV	Armored personnel carrier—armored infantry fighting vehicle
ARTY	Artillery
ATGW	Anti-tank guided weapon
CBO	Congressional Budget Office
CFE	Conventional Armed Forces in Europe
CIS	Commonwealth of Independent States
CJTF	Combined Joint Task Force
CSBM	Confidence- and security-building measure
DPSS	Designated Permanent Storage Site
ECE	Eastern and Central European
EFTA	European Free Trade Association
EU	European Union

Helo	Helicopter
HET	Heavy equipment transporter
HVO	Defense council (Croat)
INF	Intermediate Nuclear Forces
JNA	Yugoslav National Army
MBT	Main battle tank
Mort	Mortar
MRL	Multiple rocket launcher
NACC	North Atlantic Cooperation Council
NATO	North Atlantic Treaty Organization
OECD	Organization for Economic Cooperation and Development
OOV	Object of verification
OSCE	Organization for Security and Cooperation in Europe
OSD	Office of the Secretary of Defense
PfP	Partnership for Peace
SAM	Surface-to-air missile
SHAPE	Supreme Headquarters, Allied Powers, Europe
START	Strategic Arms Reduction Treaty
TLE	Treaty-limited equipment
VCC	Verification Coordinating Commission
WEU	West European Union

Chapter One

INTRODUCTION

It is difficult, in 1997, to conceive of European conventional arms control as a major issue; after all, the Cold War is over. But it remains important for the 30 signatories of the Conventional Forces in Europe (CFE) Treaty and the former warring parties in Bosnia. For CFE parties, the treaty has come to be seen as the baseline for European stability and a bellwether for the health of European international relations.[1] For the Balkan contestants, conventional arms control contributed directly to whatever stability they enjoy under the Dayton Agreement.[2] Seven years after the conclusion of the Cold War, Europe remains engaged in conventional arms control.

Conventional arms control, as distinct from its objectives, involves negotiated agreements and unilateral declarations of reductions, limitations, and constraints that states accept on their conventional forces in the hope of reducing the expense of defense, the chances of war, and the destructiveness of war if peace fails. For the past two years, RAND has been supporting the Office of Non-Nuclear Arms Control within the Office of the Secretary of Defense (OSD) in its role in conventional arms control. When the effort began, the emphasis was on preparing the United States for the May 1996 CFE Review Conference, which reviewed implementation and compliance with

---

[1]See Michael Mandelbaum, *The Dawn of Peace in Europe,* New York: Twentieth Century Fund, 1996, who suggests that the treaty is a key element of a new European security regime.

[2]Formally, the title is the General Framework Agreement for Peace in Bosnia and Herzegovina. It is also known as the Dayton Accords. Salient arms control provisions are in Annex 1-B, "Agreement on Regional Stabilization."

the treaty's provisions. After the review conference, the project refocused on anticipating the "Adaptation Talks," which opened in January 1997 and seek to adapt the treaty to the conditions of post–Cold War Europe. At this writing, those talks are expected to continue for a total of 18 months.

Supporting U.S. preparation first for the CFE Review Conference and subsequently for the Adaptation Talks has offered the opportunity to explore a wide variety of issues, ranging from the broadest policy questions to the technical details of evaluating the pros and cons of individual arms control propositions. This experience has produced some insights and observations that are potentially valuable for the ongoing adaptation process and perhaps for other instances of international negotiations—hence, this report.

The report describes the main activities and involvements of the project. It features two principal chapters, one dealing with the big questions about the future of CFE and one that describes more-technical, nuts-and-bolts, and methodological details. A final chapter suggests what can be learned from the past two years of arms control support and offers some brief recommendations for the United States' conventional arms control agenda.

Chapter Two begins with the big questions. It discusses how a specific set of issues came to be, and recaps the project's analysis and conclusions about each one. As always, some of the recommendations were overtaken by the dynamics of the negotiations, the need to build consensus, the need to compromise, and similar forces. The basic analysis, however, remains salient because it illustrates how arms control can help with contemporary security issues.

Chapter Three introduces other, more technical matters. It emphasizes the analytic tools used to illuminate some of the choices confronting the United States, and the processes used to consider alternative arms control measures. In addition to the analytic instruments themselves, this chapter considers which tools are flexible and fleet enough (that is, could yield results in time) to support questions under discussion in ongoing negotiations.

Because the Adaptation Talks are just gathering steam, it is extremely difficult to anticipate the results in any useful detail. Instead, Chapter Four consolidates the general lessons from the project and offers some recommendations about future U.S. policy for conventional arms control in Europe.

Chapter Two

"TWENTY QUESTIONS"

The project began with two workshops in RAND's Washington, DC, offices. The sessions brought together members of the extended interagency arms control family, including the Departments of Defense, Commerce, State, the intelligence community, the Arms Control and Disarmament Agency, former negotiators, and other officials. Mining the collective experience of those in attendance provided an initial list of major questions that would have to be addressed as the United States moved toward the CFE Review Conference. In addition, the ongoing interaction with the Office of Non-Nuclear Arms Control kept the project members apprised as issues changed and the debate advanced within the interagency arms control family. This chapter recounts the questions of the day and the project's answers to them.

THE BIG ISSUES

The major questions examined in the project fell into two categories, those dealing with the review conference and later negotiations and those dealing with the conditions expected in the European security environment.

Questions about the negotiations focused on the United States' preferences and options for conducting the CFE Review Conference and the Adaptation Talks were

- How much latitude is there for changing the treaty without requiring new ratification action by members' parliaments?

- How advisable is changing the treaty in the first place?
- Who are the potential troublemakers?
- What outcome does the United States seek from the Review Conference and the Adaptation Talks?

Issues about the security environment were intended to get a sense of CFE's relative importance in managing regional peace and stability. The United States would need a clear notion of what was at stake, what was important, and what was less so if it was to succeed in the talks. The questions were

- What challenges does European security face?
- How effective is the rest of the regional security architecture at addressing the area's problems?
- To what degree does CFE address current security concerns and contribute to stability in Europe?
- How will CFE and NATO adaptation interact?

The following pages offer answers.

Answers About Negotiations Details

How much latitude is there for changing the treaty without requiring new ratification action by members' parliaments?

Over the course of the project, expectations about the need and advisability of a new ratification process have changed significantly. During the early workshops, many participants placed a premium on advancing proposals that were modest enough in scope and effect not to require parliamentary approval. Gradually, as work on the European issues discussed below advanced, numerous participants concluded that the early, limited propositions would not produce a treaty demonstrably more capable of managing the new security agenda than the present CFE Treaty.

The initial inclination to discard proposed treaty adjustments likely to require ratification has moderated, especially after the May 1996 flank agreement, which in effect reduced the size of the treaty's flank zone, was referred to the Senate. Some observers, more confident that larger changes can be accomplished as "adjustments" not requiring parliamentary approval, have also become less skeptical of ambitious proposals.

Much of the project effort has been to explore various treaty adjustments and to understand their effects on European security. Greater exploration has prompted more-open consideration of major revisions that would require ratification. That the ongoing Adaptation Talks will produce a new treaty requiring formal ratification seems highly likely, given the willingness on the part of many of the participants to contemplate proposals ambitious enough to require parliamentary endorsement.

How advisable is changing the treaty in the first place?

During the preparations for the CFE Review Conference, few thought that adjusting the treaty was unwise. However, one widely held view was that the West had a "good deal" in the current treaty and should not be eager to give it up. Another camp asserted that the CFE Treaty still had substantial value in its present form, and the United States' objective should be simply to preserve the treaty.

As the debate matured and analysis suggested that there was less congruence between the CFE Treaty and the current European security equation than many thought, attitudes began to moderate. When the Review Conference concluded with an agreement for follow-on talks, it became clear that many of the signatories had more-elaborate agendas for modernizing CFE. The project conducted interviews with delegations from key arms control actors, including Russia, Ukraine, Belarus, Hungary, the United Kingdom, and the Netherlands.[1] Many of those interviewed made compelling cases for more-radical adjustments to CFE than had previously been considered. The Russians and Ukrainians cited internal security requirements as their reasons for pursuing adjustment, but other reasons were proffered too, most of which sought to address the changing nature of relations among European states.

[1]Author's interviews were conducted in Vienna on September 9 and 10, 1996, among the national delegations to the Joint Consultative Group.

Most observers, especially in light of the July 23, 1997, "Certain Basic Elements for Treaty Adaptation" agreement, which sets the stage for the remainder of the Adaptation Talks, probably agree at this point that modifying the CFE Treaty a fair amount is acceptable, if not advisable. Many at present are eager to recast the treaty for greater utility for Europe's current security circumstances.

Who are the potential troublemakers?

Early on, most knowledgeable observers answered "Russia." However, Russia arguably proved somewhat reasonable at the CFE Review Conference, asking for and receiving a deal on the flanks, reducing the amount of Russian territory subject to flank-zone restrictions, thereby increasing Moscow's flexibility to deploy treaty-limited equipment (TLE) to manage its internal security concerns more effectively.[2] How Moscow behaves during the remaining Adaptation Talks probably has as much to do with how it views the outcome of the March 19, 1997, U.S.–Russian Summit in Helsinki and NATO's Madrid Summit in July, where the issue of alliance expansion was decided. Madrid and the NATO-Russia Charter, the document codifying the new relationship between Moscow and the alliance, cite CFE adaptation as a key process for allaying fears about remaining security issues. Moscow's expectations for CFE adjustments are probably significant, which could make Moscow a more demanding and adversarial interlocutor in the remaining Adaptation Talks.

Russia is not the sole troublemaker, however. The flank states, Norway and Turkey, have long been worried that the United States might compromise their local security interests in favor of Moscow's. If they conclude that the NATO position offers insufficient protection for their equities, they could break with the alliance and offer national positions, thus complicating negotiations. The GUAM states, Georgia, Ukraine, Azerbaijan, and Moldova, want to ensure that adaptation prevents Russia from stationing forces on their territory.

---

[2]Treaty on Conventional Armed Forces in Europe (CFE) Review Conference 1996 communiqué, May 16, 1996 [Arms Control and Disarmament Agency (ACDA) Homepage, http://www.acda.gov/]. See also the U.S. OSCE (Organization for Security and Cooperation in Europe) Delegation paper "Adaptation of the CFE Treaty," Lisbon, December 2 , 1996 (ACDA Homepage, http://www.acda.gov/).

Such a provision might interfere with NATO's ability to post modest forces within the territory of its future new members.

The Adaptation Talks might become an alternative venue for states whose attempts at exerting influence have been frustrated elsewhere. For example, if France becomes frustrated in its campaign to acquire a Major NATO Command, Paris could exert its influence in CFE by demanding inclusion of some of its earlier pan-European security treaty ideas.[3] Never widely embraced, these or other maverick interventions could complicate the agenda at the Adaptation Talks.

What outcome does the United States seek from the Review Conference and Adaptation Talks?

The project postulated a comprehensive vision of Europe. In this vision, the U.S. objective is encouraging peace and stability in order to promote conditions favorable to U.S. investment and commerce. The vision has three elements: military security, economics, and human rights.

Military Security. In the military security portion, the United States encourages the Europeans to cooperate to raise forces for crisis management and force projection. Toward this end, Washington promotes common standards among NATO, the West European Union (WEU), and the Commonwealth of Independent States (CIS) for communications equipment so that member states will have compatible tactical and strategic communications for their rapid-reaction forces. The United States also promotes the notion that each state in Europe should develop the ability to deploy a brigade-sized force and sustain it for 120 days under any of the principal security organizations or as a member of an ad hoc group. Through bilateral military-to-military staff talks, NATO interaction, and the Partnership for Peace, the United States assures a minimum agreed-upon level of interoperability: common doctrine and military practices for essential operations. The United States encourages subregional cooperation and consultations on the assumption that neighbor states have similar security concerns.

---

[3]The thrust of many of these ideas was to relegate NATO to a less prominent role and to promote security discussions in broader fora, such as the OSCE.

The United States' approach to improving European institutions finds Washington supporting many of those institutions and reinforcing success. Thus, if the OSCE (Organization for Security and Cooperation in Europe) Conflict Prevention Center proves unusually effective as a forum for soliciting participants in crisis-monitoring missions, the United States would invest more effort in OSCE. Likewise, if NATO proves very effective at managing security issues, the United States would redouble its efforts there.

Washington adopts a different tack, a major departure from its traditional leading role, in the conventional arms control agreements of the region, leaving them to other powers to break or amend. If and when the United States must engage in additional European arms control talks, it advocates subregional pacts to address local problems, discourages further pan-European treaties, and, when pressed, often prefers new counter-concentration and counter-surprise[4] confidence- and security-building measures (CSBMs) over expensive equipment-reduction deals.

Economics. In the economic portion of the U.S. vision, the United States seeks access to markets and economic stability that will make Europe a sound, profitable site for American investment and business. Thus, Washington wants European states with stable, convertible currencies; marketable goods of their own; and growing gross domestic products—states that will consume increasing numbers and quantities of U.S. products. To help realize this part of its vision, the United States seeks to persuade the European Union (EU) to undertake more-meaningful trade with the Eastern and Central Europeans; promotes retraining of Eastern and Central European officials in capitalist money, banking, and public administration courses offered at U.S. and European universities and similar institutions; and promotes loans and credits throughout the region. The United States encourages the Organization for Economic Cooperation and Development (OECD), the World Bank, and others to adopt a coordinated aid policy that focuses first on the healthiest, least-dysfunctional developing economies in the hope that they will

[4] *Counter-concentration measures* prevent massing of military forces for offensive action. *Counter-surprise measures* provide insight into military activities. Such insight would limit a state's ability to mount an attack against another state with little or no warning.

become fully productive quickly and thus be able to contribute to the rejuvenation of the most seriously ill economies, which, presumably, will require proportionately more help. The United States discourages *disproportionate spending*—spending at levels that would jeopardize overall economic health and performance—on military modernization.

Human Rights. The human rights portion of the vision seeks to reinforce a secure business environment by promoting the *rule of law*—agreement to a basic set of human rights including protections for minorities—and an agreed-upon approach to citizenship based on place of birth and loyalty rather than on blood ties or ethnicity. This approach will support a secure environment for investment and business while tempering incentives for migration and insurgency.

Toward this vision, Washington supports the various efforts of the OSCE and others. As with military security, in human rights the United States supports numerous efforts (e.g., reasonable citizenship standards, regular elections) and reinforces the successful ones.

This vision of Europe and the U.S. role in it maintains the *security redoubt*—CFE and NATO in their classic roles—as a hedge against the return of the "bad old days," but is more focused on programs to address the immediate sources of potential trouble in Europe. Because the United States pursues its agenda in many fora, opportunities abound for NATO, EU, OSCE, OECD, and other organizations to meet, consult, and debate. There is potentially enough activity to offset fears about any organization's becoming moribund. The United States' decision to work in all fora and reinforce success will thus have the effect of keeping all organizations "warm," like an assembly line fully geared up but not yet in production.

Answers About European Security

What challenges does European security face?

Security and stability are closely linked in international affairs. Security is about safety, and stability is about consistency in the international order. Without stability, security is elusive.

Fundamentally, *security* is about the safety of one's life, family, and property. During the Cold War, the Soviet Union posed the largest threat to most Western Europeans. Today, the threats are more numerous. Crime, terrorism, religious or ethnic repression, local wars, economic recession, and unemployment have their origins closer to home. Russia is unlikely to be able to stage large-scale military operations for some time, and any program to make her capable would offer many indicators and ample warning. NATO and the European security architecture remain in place to counter and deter any return to confrontation.

Security rests in part on stability. *Stability in international relations* means an absence of challenges to the international system. During the Cold War, Soviet expansionism, fueled in part by Marxism-Leninism, was the principal threat to the status quo. In 1990, the CFE Treaty provided for military stability and reduced the prospects of warfare further by producing a balance of forces among competing blocs of states so that no side could gain surprise and generate an overwhelming advantage in military forces. In 1997, the old threats to stability have vanished. Both the Soviet Union and its ideological basis are gone.

Now, threats to stability come from the former Yugoslavia, Bosnia, Albania, and from beyond Europe. Anti-Western actors, convinced that they have suffered as a result of European actions, including colonialism and cultural imperialism, are eager to punish. Refugees and immigrants from non-European cultures produce new pressures for change. National economic well-being and unemployment are contingent on the performance of the international markets, multinational corporations, and other forces that now lie beyond the sovereign power of most European states. Stability is becoming more vulnerable to forces and phenomena generated by the interdependence of states and the free flow of people and ideas across the borders of post-Maastricht Europe.

NATO and CFE have only indirect effects on the current threats to security and stability. CFE in its present form may continue to regulate the upper limits of military forces and perform its other tasks, such as military transparency and compliance verification. But it is increasingly irrelevant among nations that are unlikely to fight each other and that, if they do fight, will fight with much smaller armies than CFE allows, since they cannot afford larger forces. Moreover, CFE does not fully address all the new threats to stability.

Neither CFE nor NATO addresses the challenges to security and stability that seem most likely: threats emerging at and beyond Europe's borders. Together, CFE and NATO form a security redoubt in the event that Europe confronts Cold War–like threats again. The CFE Treaty and NATO safeguard core European and U.S. security interests by ensuring that Europe cannot be dominated by a hostile power. Beyond this bedrock-level interest in maintaining Europe's freedom, the United States has considerable policy latitude. Given the abundant security architecture, the quality of security and stability in Europe should not be extremely sensitive to most adjustments to CFE.

How effective is the rest of the regional security architecture at addressing the area's problems?

Pluses. The Continent began creating new organizations and structures in which to couch its security and well-being shortly after the Second World War. The year 1949 witnessed the birth of the Council for Mutual Economic Assistance, NATO, and the Council of Europe. The Nordic Council, Western European Union, Warsaw Treaty Organization, Benelux Economic Union, Organization for Economic Cooperation and Development, the European Free Trade Association (EFTA), European Community (later the European Union), Conference on Security and Cooperation in Europe (CSCE; later OSCE), and others appeared as the decades rolled by. Collectively, these organizations sought to provide economic growth and prosperity, along with stability and international security to the region.

In addition to organizations, a host of agreements on arms control and confidence- and security-building measures emerged from the late Cold War era. Nuclear arms control provided Strategic Arms Limitation Talks and, later, in 1991 and 1993, two Strategic Arms Reduction Treaties (STARTs). These pacts were complemented by an agreement to destroy intermediate nuclear forces (INF) in Europe and by President George Bush's Presidential Nuclear Initiative withdrawing tactical nuclear weapons, a move reciprocated by the Soviet Union. Chemical and biological weapons conventions, a mutual aerial-observation regime known as Open Skies, and an agreement on ceilings for military manpower (CFE1A) filled in the equation. Together with the CFE Treaty, these agreements and organizations constituted the European security architecture. Some of the arms control agreements failed to attain ratification in all capitals, and some remain to be implemented. Nevertheless, they have prompted cooperative discussions that have been sustained through some of the worst tensions of the East-West confrontation years.

Among the organizations incorporating the most European members, the OSCE—the principal forum for arms control—is arguably the most prominent. Over the years, its members have crafted standards for human rights, crisis management, and the handling of a host of other problems. Its 1990 Charter of Paris for a New Europe created the first pan-European structure for regular consultations among governments and the first fully cooperative institutions to deal with European security problems. The member states agreed to a Council of Foreign Ministers to meet at least once yearly, a Committee of Senior Officials to meet more frequently to carry out regular business among the members, a full-time Secretariat to organize the OSCE's business, a Crisis Prevention Center, and an Office of Free Elections to oversee and safeguard nascent human rights. The OSCE Budapest Document of 1994 provided an astonishing number of agreements, including arrangements for a global exchange of military information, a set of principles governing arms transfers, stabilizing measures for localized crisis situations, principles for governing nonproliferation, controls on chemical-biological weapons, and a control regime for missile technology. The document's most ambitious accomplishment was a set of principles for a Common and Comprehensive Security Model for Europe for the Twenty-First Century.[5]

Laudatory though the OSCE and the rest of the security architecture may be, few have much faith in these structures. With 53 members generally governed by rules of consensus, the OSCE has little leverage over individual members; it will likely find itself hamstrung on any truly contentious issue. In addition, many Europeans fear duplicating capabilities held elsewhere. Thus, while the OSCE has gained a Secretariat and a few other standing bodies in recent years, it still lacks the organization to function effectively or to oversee large activities. Few in the West would accept it as the key organ for guaranteeing regional cooperation and security.

---

[5]Section VIII, Budapest Document, December 21, 1994.

Nor are many of the institutions in the security architecture perceived to have much moral weight. Only NATO stands as a symbol of real security and real collective self-defense. The EU and other organizations offer obvious economic advantages; however, when classical security issues head the agenda, only NATO appears to most Western and Eastern and Central European (ECE) states as a credible organization for safeguarding the region's peace and security.

Shortcomings. Europe's current security posture is not yet fully coherent. Fortunately, although some potential foes are waiting in the wings, the Continent faces no immediate, major challenges to the status quo. Many European states have experienced the benefits of cooperation directly, as evidenced by the gradual upturns of their economic fortunes. The 53 members of the OSCE have agreed to an extraordinary number of localized regimes and procedures that should improve security and cooperation.

Europe is secure, yet remains afraid. Despite the elaborate dimensions of the security architecture, few trust the institutions having the most-comprehensive membership, including OSCE, to provide reliable security. The Balkan conflict, reined in by the Dayton Accords and the presence of NATO peacekeepers, remains unresolved and could reignite with little warning. Although some 30,644 tanks, 54,181 armored combat vehicles, 30,426 artillery pieces, 1,613 attack helicopters, and 6,548 combat aircraft remain within the European military forces of the region,[6] these forces are ill-prepared and poorly organized to protect Europe against the potential internal conflicts or the external foes lurking around her periphery.

NATO, the only element of the security architecture with the military means and potential to handle contemporary European security problems, is also in a period of transition in which it is less than ideally suited for current challenges. Its slow, deliberate approach has produced limited progress, but the approach remains inchoate: Its member states have not yet fully embraced a strategic concept

---

[6]These numbers reflect the principal ground and air force holdings of Belgium, Denmark, France, Germany, Greece, Italy, the Netherlands, Norway, Portugal, Spain, Turkey, the United Kingdom, Albania, Austria, Belarus, Bulgaria, the Czech Republic, Finland, Hungary, Poland, Romania, Slovakia, Sweden, and Ukraine. Data are from International Institute for Strategic Studies (IISS), *The Military Balance*, London: Brassey's, 1996.

appropriate to the present security environment. Despite some progress, its military forces generally remain organized for direct self-defense of their territory and are constantly scrutinized for further downsizing to garner greater savings for their governments.

To what degree does CFE address current security concerns and contribute to stability in Europe?

CFE's Cold War Pluses. CFE was well designed for the bipolar confrontation of the Cold War. The treaty controlled the weapons thought to be most militarily significant for offensive operations in any NATO–Warsaw Pact war. It caused both blocs of states to reduce their treaty-limited equipment to equal amounts: 20,000 tanks, 20,000 artillery pieces, 30,000 armored combat vehicles, 6,800 combat aircraft, and 2,000 attack helicopters. It caused the participants to destroy some 50,000 pieces of TLE and to carry out some 2,700 inspections. The subzone structure of the treaty, consisting of nested zones and a flank region, limited each bloc's ability to concentrate its TLE and produce the overwhelming force ratios desirable for offensive operations. Stability resulted, since neither side could generate a surprise attack while observing the limitations of the treaty.

To ensure that signatories complied with the treaty's provisions, intrusive verification and monitoring provisions were included. Moreover, a series of CSBMs codified in the Vienna Document[7] required states to invite observers to exercises above a certain size and to notify the other signatories 42 days in advance of exercises involving certain numbers of troops, and limited exercises involving 40,000 troops or 900 battle tanks to once every two years. These provisions dispelled suspicion about states' motives and activities and contributed to the prevention of surprise attack.

CFE's Post–Cold War Shortcomings. Given the transformed security landscape in Europe, CFE as currently configured is much less central to Europe's problems. The treaty does little to control strife between two European states, because the subzones that were in-

[7]The Vienna Document was first negotiated among the full membership of the OSCE in 1990. It was last updated in 1994, providing a set of politically binding CSBMs; transparency measures, which allow observers to understand a country's military organization and capabilities; and measures to dispel military ambiguity.

tended to prevent the massing of NATO and Warsaw Pact forces along the Inter-German Boundary do not prevent smaller concentrations in central Europe between neighbors with unsettled grievances. The flank provisions are a source of irritation between Russia and the West, since Moscow insists that its internal security requirements demand more TLE in the flank regions than the treaty allows. The other flank states, Turkey and Norway, have been adamant about Russian compliance.

The Vienna Document CSBMs requiring advance notice for exercises and observers under some circumstances still have important utility. But they, too, were designed to prevent the surprise massing of very large forces, not the smaller forces that threaten ethnic or interstate conflict in today's Europe. In a similar way, the CFE inspection system and quotas seem less adequate today because more states are interested in inspecting Russia and their immediate neighbors than they were before. In short, the inspection regime and counter-surprise features of the current CFE Treaty and Vienna Document do not adequately address current concerns.

With the demise of the Warsaw Pact, the symmetry of the two blocs is gone. CFE does not now much constrain NATO. Russia could insist that any former Eastern states that joined NATO would have to have their TLE counted under NATO's total, but NATO now has enough headroom[8] to incorporate a few new members. Nor are there provisions that would limit Combined and Joint Task Forces (CJTFs) or similar multinational formations. Russia is already calling for new measures for controlling alliances and for placing limitations on the forces of states in political-military structures. The treaty's only measure controlling potentially dominant states or the regional distribution of power is the Sufficiency Rule, which simply limits the amount of TLE any one state can have to roughly one-third of the totals for the *area of application*—the area from the North Atlantic to the Ural Mountains.

Although CFE managed a major concern during the Cold War, the fear of surprise conventional attack, the treaty clearly plays a more limited role given the post–Cold War security agenda. At best, many

---

[8]Unused entitlements to TLE.

of its measures have served as models for later security solutions. For example, many of the Dayton Accords' arms control features reflect the influence of CFE: zones of separation similar to CFE subzones, treaty-limited equipment, and monitoring.

In any case, the direct contribution to European security that CFE currently makes is much smaller than its contribution would have been during the Cold War era.

How will CFE and NATO adaptation interact?

NATO-CFE Interaction. Russia's reactions certainly will be key. However, the interaction between the two treaties themselves is also noteworthy. The effects of NATO on some of the key attributes of CFE are best appreciated in terms of those measures of the NATO Treaty that produce stability, lower force levels, limit offensive capability, and reduce capability for surprise attack.

We now consider how NATO enlargement, development of a new strategic concept for the alliance, a revised command structure, and NATO-CFE interaction might influence each other.

The Effects of NATO Enlargement. NATO enlargement would have a negative effect on CFE. Russia has made its objections clear, although its probable course of action if the allies ignore Moscow's warnings remains to be seen. In any case, Russia has significant expectations that her concerns about NATO will be addressed in CFE. Trouble with Russia over CFE could presage a general decline in European security that could have broader consequences. Russia could demand substantial changes to CFE and relief from key provisions or, under extreme circumstances, could scrap the treaty. The quality of security in Europe would be damaged: Stability would suffer because of the Russian perception that East-West competition had been restarted. Moscow would probably assess NATO as an expansionist, anti-Russian entity and determine to contain it. Despite her paucity of resources, Russia would do what she could to undermine further regional stability, to shore up her frontiers, and to restore her ability to confront the West. This competition would create new incentives for Moscow to resist the expansion of Western influence and to offset NATO expansion with initiatives of her own: perhaps limited arms racing and alignment with other extra-European, anti-Western forces, for example. NATO enlargement would thus be bad not only for the CFE Treaty but for European stability as a whole.

Effect of a New Strategic Concept. One of the first elements to emerge from the allies' ongoing discussions of strategic concept has been the CJTF. Because the alliance will probably determine that the preponderance of its military activities will be CJTF expeditionary operations, NATO enlargement probably need only provide agreed-upon tasks, conditions, and standards for the basic repertoire of military missions, a calendar of alliance training activities and exercises, and a scheme for assigning and earmarking units for CJTF duty.[9] The new member states will meet their obligations with relatively small contributions of reaction forces, adoption of NATO tactical standards and practices, and regular participation in alliance training events.

The new concept would improve power projection and expand NATO's capability to employ force across the alliance and also beyond the territory of its members. The new concept would also improve collective offensive capabilities while keeping them modest. The counter-surprise features of the treaty remain intact. Most allies maintain relatively small standing forces for limited, crisis response–type missions. The preponderance of allied forces remains reserve units that require mobilization and predeployment training before they can undertake military operations. This force posture—reliance on large reserves—ensures unmistakable indicators and warning of large-scale military activity, and thus reduces the likelihood that NATO could surprise an adversary.

Changed Alliance Structure. A reorganized alliance command structure—redesign of Major NATO Commands and reapportionment of those commands to be headed by representatives of key allies—will eventually be the operational, practical expression of NATO's new strategic role and concept. The assessment of new structure parallels that of the new strategic concept. Only the most-ambitious organizational schemes that would provide a Major Command for power projection are likely to injure CFE's key attributes. The more-modest

---

[9]NATO is in the process of developing Combined Joint Task Forces as the principal instrument for military action other than collective self-defense. For a detailed discussion of the process, see Charles Barry, "NATO's Combined Joint Task Forces in Theory and Practice," *Survival*, Vol. 38, No. 1, Spring 1996, pp. 81–98.

proposals for corps-sized CJTFs subordinated to Major NATO Commands seem fairly benign.

Interaction of NATO and CFE Adaptation. The interplay of NATO and CFE produces mixed results. The greatest difficulty is uncertainty about the specific proposals for modifying CFE. For example, revising the bloc structure of the treaty would be fine insofar as NATO modernization goes and as long as there were no penalty for NATO enlargement. But certain proposals, such as the suggestion advanced by Russia that the military forces of states that are members of "political-military structures" (i.e., alliances) should be subject to tighter constraints than states that are not, might try to limit the alliance's size and influence.

Similar uncertainties surround the TLE-redistribution proposals. Any approach that would reduce NATO's aggregate allocations or that would produce a ceiling that the alliance could not exceed, even when accepting new members, would be potentially dangerous.

The potential trouble with new or modified TLE levels and CSBMs lies in their ability to constrain NATO deployments, stationing, and operations. Russia's objections to NATO forces being stationed in Central Europe are well known. Moscow or other capitals might introduce proposals seeking to capture NATO by placing limits on multinational operations (e.g., no more than so many troops from so many countries) in such a way that wide participation in alliance operations and training activities would be undermined. This capturing would constitute a different problem from Russia's insisting that all NATO TLE fit within the Western group-of-states entitlement, where some headroom exists. Such proposals could impinge upon the alliance's options for its command structure and strategic concept.

Whatever the merits of NATO elsewhere in the European security equation, the main emphases of alliance modernization (particularly enlarged membership) are harmful to the main attributes of CFE. In addition, some of the proposals for CFE adaptation offer inroads for measures that could constrain NATO-modernization options. For example, Russia has suggested new measures that would apply only to members of alliances.

DIFFERENT PROBLEMS, DIFFERENT TOOLS

The Europe of today and the Europe of tomorrow face problems different from those CFE was designed to control. Indeed, regionwide arms control in the present circumstances does not seem very promising as a major policy instrument. Many states in Europe facing internal pressures, friction with neighbors, and threats from beyond Europe's frontiers need other, more highly customized tools than CFE affords. But what are these tools and how should they be selected? Chapter Three suggests some answers.

Chapter Three

ANALYTIC EXCURSIONS

The project explored three general groups of questions.

First, how can provisions of the current CFE Treaty be modified to conform to the signatories' new security preferences? If the current, bloc-to-bloc structure of the treaty were abandoned in favor of national limits, how would all the other treaty provisions, also predicated upon blocs, be adjusted? This group of questions dealt with matters including how to derive new TLE entitlements and how to calculate liability for inspections.

The second group of questions examined options for new measures. If the CFE Treaty were to be updated to contribute more directly to European security, what new measures might be possible? How could they be tested to gauge their comparative effectiveness? For this group of questions, we researched the possibilities of new categories of treaty-limited equipment, subregional measures, new counter-concentration measures, and new counter-surprise measures.

The third group of questions focused on modeling as a means to help the project team think about new ways to recast the treaty for contemporary Europe. Since Chapter Two has already examined the prospects for new security measures in existing arms control categories (e.g., transparency), this effort approached the problem differently by confronting the CFE with new circumstances and then trying to figure out what modifications would be necessary to make CFE more effective under the conditions posited.

This chapter revisits the work on each group of questions.

GROUP ONE: MODIFYING CURRENT TREATY PROVISIONS

In initial project efforts, we emphasized modifying four existing elements of the CFE Treaty: TLE entitlements, verification and inspection, transparency, and counter-concentration of forces.

TLE Entitlements

The original basis for TLE entitlements—that each of two groups of states was entitled to 20,000 tanks, 30,000 armored combat vehicles (ACVs), 20,000 pieces of artillery, 6,800 combat aircraft, and 2,000 attack helicopters, which they in turn allocated among their member states—is untenable because the Eastern group of states no longer exists. Moreover, from NATO's perspective, perpetuation of the group-of-states structure is dangerous: Russia might argue (contrary to current treaty language) that states wishing to join the alliance should be forced to adhere to the existing group-of-states cap on TLE, thus restricting the military capability of the alliance as a whole. As a practical matter, TLE entitlements have devolved to national ceilings and national reporting, verification, inspection, and reduction responsibilities. The question is how to convert these circumstances so that the treaty becomes fully state-based instead of based on groups of states. We propose two answers: (1) adopt current allocations or (2) renegotiate individual-state ceilings.

Adopt Current Allocations. Under this proposal, the 30 CFE signatories would wait until Russia renegotiates TLE allocations for the flanks[1] with the Tashkent participants (other states of the former Soviet Union) and then agree to adopt their respective current allocations as national ceilings. The benefits of this approach are obvious: The approach is simple, does not revisit sufficiency rules or other potentially contentious issues, and leaves the Atlantic-to-the-Urals zone with essentially the same TLE entitlements as those in place today. This approach could be easily derailed if Moscow insists on some bloc-limitation scheme, however. For example, Russia will certainly recognize the potential to capture NATO expansion under a bloc-TLE cap and will probably be reluctant to give up this leverage unless the NATO signatories to the treaty somehow reassure Moscow

---

[1] An outcome of the May 1996 CFE Review Conference.

that Russia has nothing to fear and perhaps something to gain from alliance expansion.

Renegotiate Individual-State Ceilings. In this approach, the 30 CFE signatories would renegotiate their TLE entitlements subject to the following guidelines. First, current sufficiency rules apply. Second, the total TLE allowed in the area of application would remain limited to the treaty's present maximums: 40,000 tanks, 60,000 armored combat vehicles, 40,000 artillery pieces, 13,600 combat aircraft, and 4,000 attack helicopters within the treaty's area of application. Third, each state's individual TLE ceiling must be consistent with the TLE-to-space and similar ratios of the participants in general. The project team made several explorations to understand current national TLE holdings and to look for potential norms or similarities among countries. First steps included examining TLE-to-space ratios on a regional and more local basis. Figure 3.1 illustrates this excursion.

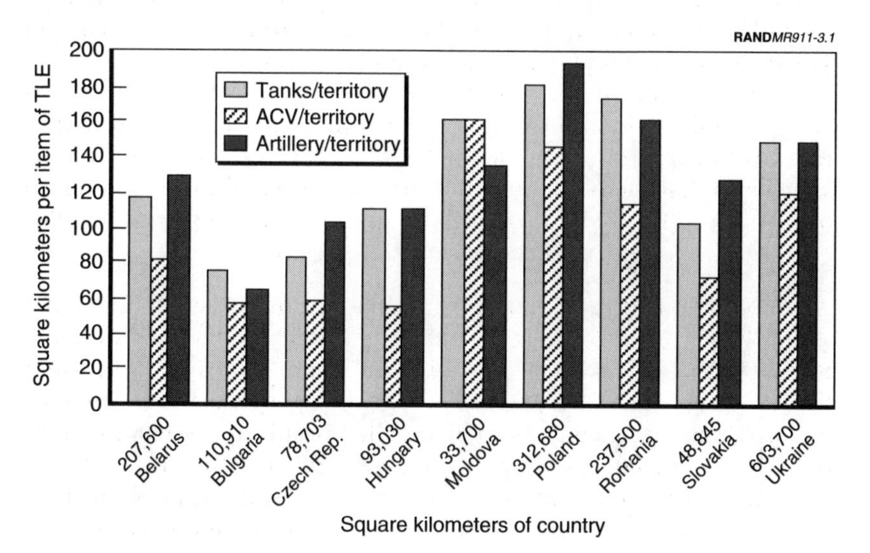

SOURCE: Annual information-exchange declarations, January 1995, and the *Discovery World Atlas*, Maplewood, NJ: Hammond and Co., 1988.

Figure 3.1—Comparative Eastern and Central European TLE-to-Space Ratios

We assumed that there is a relationship between a country's need for TLE and the amount of territory it has to defend. If the parties could agree on a norm—perhaps the average number of square kilometers per item of TLE plus-or-minus some percentage—there might be a rational basis for calculating individual-state entitlements. This approach would let the parties negotiate a formula for TLE entitlements that would take the signatories to similar, or lower, levels than those currently in force, depending on the wishes of the individual member states, but driven by a common conception that a certain amount of TLE is necessary for self-defense within a certain amount of territory. However, as the figure illustrates, the problem is that the density of equipment varies by almost a factor of 3 among the states. Thus, agreeing on a norm would probably prove very difficult, since the more heavily equipped states would face large reductions.

Next, in a similar effort, we compared TLE with national-boundary length, shown in Figure 3.2, hypothesizing that if some states had longer, more-convoluted frontiers to protect, they might legitimately be entitled to larger TLE holdings. This figure seemed potentially il-

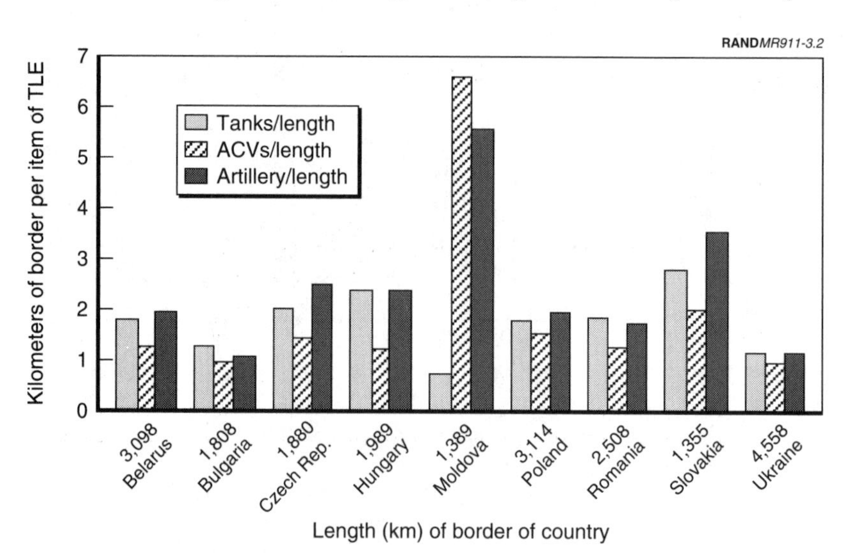

SOURCE: Annual information-exchange declarations, January 1995, and the *Discovery World Atlas*, Maplewood, NJ: Hammond and Co., 1988.

Figure 3.2—Comparative Eastern and Central European TLE-to-Borders Ratios

luminating because, except for Moldova, most states' ratios were somewhat closer to each other's than the TLE-to-space ratios. Still, a useful norm eluded us.

Figure 3.3 illustrates our third attempt to find useful, instructive ratios. We posited that countries having more kilometers of railroad track and paved roads, plus more military-quality airfields, can move TLE more quickly than those with less of such infrastructure. Infrastructure therefore became the basis for inflating or deflating TLE entitlements. Although interesting, the assumption seemed too complicated to implement, because such an adjusting factor would obviously be very sensitive to small differences in key infrastructures such as military airfields, so that a small calibration error could have far-reaching effects. Imagine trying to credit the less-well-supplied states for Ukraine's relatively generous infrastructure. Just how much TLE could Ukraine be penalized, especially since she shares a border with Russia and most of the other states in Eastern and Central Europe do not? Moreover, as with the TLE-to-space ratios, the variances between individual states' infrastructures seemed just too great to facilitate finding norms.

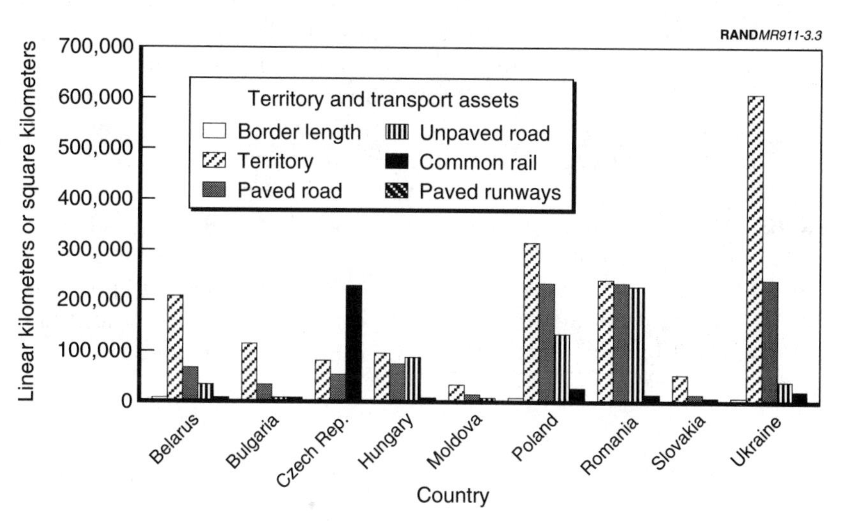

SOURCE: Annual information-exchange declarations, January 1995, and the *Discovery World Atlas*, Maplewood, NJ: Hammond and Co., 1988.

Figure 3.3—Comparative Eastern and Central European Territory and Transport Infrastructure

As a practical matter, arriving at low TLE numbers may not be too difficult. Table 3.1 summarizes current entitlements and compares them with 1997 declared holdings for selected states, indicating that many states maintain TLE levels significantly below those to which they are entitled. Current state practices of maintaining relatively low TLE levels suggest that any future renegotiation might not produce significant shifts in TLE distribution. Rather than pursue some variation of TLE-to-space ratios to find appropriate national levels, adoption of current holdings might prove satisfactory. If some states do wish to pursue substantially larger entitlements of TLE, the counter-concentration options discussed later in this chapter could be used to manage those entitlements.

Verification and Inspection

The backbone of the treaty is its verification protocol, which rests almost exclusively on the inspection regime. States derive their *passive inspection quotas*—the number of inspections they are obliged to receive from other states—from the number of objects of verification (OOVs) they declare:[2] 15 percent of their OOVs for declared site inspections and 23 percent of their OOVs for challenge inspections.[3] The *active inspection quotas*—the number of inspections a state is entitled to conduct—were determined within its group of states and represented some share of the total number of passive inspections the other group of states was obliged to accept. Thus, if the CFE Treaty moves from bloc-to-bloc to state-based obligations, some adjustment must be made to the method of determining active inspection quotas. Four solutions suggest themselves: (1) adopt current inspection quotas by acclamation, (2) base inspection quotas on costs, (3) negotiate active inspection quotas annually or biennially among the 30 signatories, or (4) supplement the inspection regime with other means.

---

[2] *Objects of verification* are military installations, storage facilities, and similar sites that hold TLE.

[3] Challenge inspections pertain to inspection of specified geographic areas that contain OOVs rather than of each individual OOV.

Table 3.1

TLE Entitlements Versus Declared Holdings

State	Tanks	AIFV	Artillery	Combat Aircraft	Attack Helicopters
Belgium	334-334	1099-678	320-316	232-166	46-46
Czech Republic	1435-952	1367-1367	767-767	230-144	50-36
Denmark	353-343	316-286	553-503	106-74	12-12
France	1306-1156	3820-3574	1292-1192	800-650	396-326
Germany	4166-3248	3446-2537	2705-2058	900-560	306-205
Greece	1735-1735	2534-2325	1878-1878	650-486	30-20
Hungary	835-797	1700-1300	840-840	180-141	108-59
Italy	1348-1283	3339-3031	1955-1932	650-516	139-132
The Netherlands	743-722	1080-610	607-448	230-181	50-12
Norway	170-170	225-199	527-246	100-74	0-0
Poland	1730-1729	2150-1442	1610-1581	460-384	130-94
Russia	6400-5541	11480-10198	6415-6011	3416-2891	890-812
Turkey	2795-2563	3120-2424	3523-2843	750-362	103-25
Ukraine	4080-4063	5050-4847	4040-3764	1090-940	330-294
United Kingdom	1015-662	3176-2411	636-436	900-624	371-289
United States	4006-1115	5372-1849	2492-612	784-624	431-126

SOURCE: January 1997 national data declarations as reported in Dorn Crawford, "Conventional Armed Forces in Europe," *CFE: A Review of Key Treaty Elements,* Washington, DC: Arms Control and Disarmament Agency, 1997.

NOTE: AIFV = armored infantry fighting vehicle.

RAND*MR911-T3.1*

Adopt Current Inspection Quotas by Acclamation. One way to avoid renegotiating the basis for inspection quotas would be to adopt the quotas currently allocated to each of the treaty signatories. This solution would maintain current rules and restrictions (e.g., no one state can use more than 50 percent of another's passive quotas) and simply adopt the current active quotas as fixed numbers of inspections. The advantage of this approach is simplicity, since it changes very little about the treaty. A potential problem exists if one or more states radically alter the number of their OOVs. The other states could then be left with active quotas inadequate to provide the current level of intrusive verification. On the other hand, if OOVs decline significantly, the number of active quotas might go unused.

Base Inspection Quotas on Costs. This solution would be to agree on a permanent percentage of quotas for all parties based on the distribution of costs associated with operating the treaty. In this arrangement, all participants would be provided with inspections proportionate to their respective shares of expenses. For example, the total 2,302 OOVs in the area of application should result in 529 challenge inspections and 345 declared site inspections, under current counting rules.[4] These inspection quotas could be viewed as "tickets" entitling the holder to conduct an inspection. The "tickets" could be distributed according to the share of expenses a state bears for operating the CFE Treaty and its standing body, the Joint Consultative Group. Other distribution plans could of course be possible.

Negotiate Active Inspection Quotas Annually or Biennially Among the 30 Signatories. Such an admittedly cumbersome approach would provide the signatories an opportunity to cooperate in addressing mutual security needs. The rules might stipulate that, until a deal is brokered, current active inspection quotas remain in effect. In this approach, the same process once carried out among the members of each group of states' parties would now take place among all 30 signatories. If the matter of active inspection quotas turns out to be more sensitive than the data thus far indicate, this solution might be the simplest way to address the problem.

But inspections are expensive,[5] and states look for ways to pool resources so that fewer inspections need to be conducted. Examination of data from the U.S. On-Site Inspection Agency (OSIA) indicates that NATO members, often in association with Eastern and Central European countries, conduct only 56–58 inspections under CFE annually. Results from these inspections are shared through the Verification Coordinating Commission (VCC) established within NATO. In addition, the confidence- and security-building mechanisms of the Vienna Document of 1994 and its predecessors also provide opportunities—albeit less intrusive ones—for states to inspect each other.

The point is that concerns about getting the number of active quotas right may not be as significant as some observers think, since there

---

[4]January 1997 national data declarations as reported in Dorn Crawford, "Conventional Armed Forces in Europe," *CFE: A Review of Key Treaty Elements*, Washington, DC: Arms Control and Disarmament Agency, 1997.

[5]The Congressional Budget Office (CBO) estimated in 1991 that annual CFE inspection costs for the United States would run between $25 and $75 million.

are other means of verifying treaty compliance and since states seem to be looking for ways to economize on the number of CFE inspections conducted. Inspection results of all types and Partnership for Peace exercise reports could be more widely shared. Ratification and implementation of the Open Skies aerial-observation regime would offer another, low-cost means of monitoring compliance with the CFE Treaty.

Supplement the Inspection Regime with Other Means. This solution would encourage Russia to ratify Open Skies. Or, given the proliferation of commercially available 1-meter-resolution satellite imagery, sensitivities and concerns about U.S. imagery may have abated enough to make it possible to share U.S. photographic imagery among the 30 signatories. Such an initiative would certainly prove a valuable and inexpensive supplement to active inspection.

Transparency

At present, transparency in CFE rests principally upon the inspection regime and the annual data exchange. Transparency in Europe is further supplemented by the provisions of the Vienna Document of 1994, which provides for notification and observation of agreed-upon activities, and by the North Atlantic Cooperation Council (NACC) and the Partnership for Peace (PfP). The NATO initiatives have produced extended staff talks, multinational peacekeeping exercises, and similar military training from one member country to another, and, with the establishment of national military liaison cells at the Supreme Headquarters, Allied Powers, Europe (SHAPE), more transparency. As a result of these Atlantic Alliance institutions, there are more mechanisms for transparency now than when CFE was signed. That said, CFE could still be adapted for greater transparency by three means: (1) expanding the current transparency measures, (2) creating a combined, long-term exercise calendar, or (3) establishing cost-benefit considerations.

Expanding the Current Transparency Measures. One approach would be to expand the current data declarations to provide more detail. Currently, nations report military formations to the brigade-regimental level under the Vienna Document of 1994. Perhaps a revised annual declaration could be negotiated that would require reporting to the battalion level, which would give insights into the combat support and combat service support units. Such a change would involve modifying existing reporting software—an expensive proposition—but would improve transparency. In a similar way, the criteria for notifying other states of exercises, removing equipment from a Designated Permanent Storage Site (DPSS), and other activities governed by CFE or the Vienna Document of 1994 might be modified to provide longer notices and to lower the threshold at which activities become notifiable. For example, instead of notifying the other signatories when an exercise will involve 300 tanks, thus alerting them to a division-level event, perhaps the threshold could be dropped to 100 tanks, resulting in notice of brigade-regimental–sized exercises.

Creating a Combined, Long-Term Exercise Calendar. The U.S. military employs a 5-year training calendar to coordinate its major multiservice and multinational events. The current CFE annual data-declaration provisions could be adjusted to include a long-term training calendar in which the 30 signatories would report any planned multinational training exercises. The benefits to transparency are obvious: Each of the signatories would have a clear picture of all the major military activities of their treaty partners.

Establishing Cost-Benefit Considerations. All of the CFE states face substantial budget pressures and operate their defense establishments under austere conditions. At some point, their spending for transparency must produce confidence. States situated in a zone of peace (e.g., Portugal, for whom the concerns of Eastern Europe must seem far removed) may want to buy less additional transparency than states in zones of fear. The flanks currently operate more transparency measures than the rest of the treaty area, suggesting that perhaps states in the principal area of application could simply subscribe to additional transparency measures, too—for a price.

In such an arrangement, two or more states would agree to provide supplemental transparency measures. The state seeking additional transparency would pay the state providing it for whatever the method might be: more-frequent reports, observers at exercises, or Open Skies–like flyovers. Moreover, as an incentive to induce states to participate, the subscriber state would also pay to provide similar transparency of itself to the other state. Each state in the treaty would thus be free to reach its own conclusions about how much additional transparency it needed.

Counter-Concentration

CFE currently provides counter-concentration through its groups of states, through Designated Permanent Storage Sites, through the geographic zones and subzones within the treaty's area of application, and through the sufficiency rules.

The two groups of states were each entitled to the same amounts of TLE, thus producing approximate military parity in which neither side could build a decisive advantage in offensive forces. DPSS limits the amount of TLE that can be deployed in active units, further reducing the immediately available military potential of both blocs of states.

The zones and subzones interfere with the massing of the available forces on the frontiers of neighbor states, requiring distribution of forces and diluting the forces available at the old line of confrontation along the Inter-German Boundary. Thus, the subzones make it more difficult to generate decisive, local offensive capability.[6]

Finally, the sufficiency rules limit any one state's TLE entitlement to a percentage of the total TLE in the area of application, preventing any single state from amassing overwhelming military capability compared with that of the other signatories.

How can the same level of counter-concentration be provided in the new strategic circumstances, where, among other things, the groups of states no longer exist, some treaty signatories want to do away with DPSS, and there are pressures to reformat the zones and subzones? We came up with four possible solutions: (1) consider a new, territorial sufficiency rule, (2) provide new, state-based counter-concentration rules, (3) consider establishing overlapping zones, or (4) agree to constraints on organization of military forces.

---

[6]The subzones were intended to manage bloc-to-bloc confrontation. Therefore, the benefits of the subzones are rather limited when dealing with state-to-state tensions.

Consider a New, Territorial Sufficiency Rule. Just as the current Sufficiency Rule precludes any state from having more than roughly one-third of the TLE in the zone of application, this new rule would provide that no state have more than 20 percent or some other agreed-upon amount of the total of each type of TLE for the zone of application on its territory, thus limiting the amount of national and foreign equipment that could be positioned on its territory. If 20 percent were the agreed-upon figure, this would mean 8,000 tanks, 8,000 artillery pieces, 12,000 ACVs, 2,720 combat aircraft, and 400 attack helicopters. As Table 3.1 indicates, such numbers would leave headroom in the ground systems adequate to accommodate stationing of some foreign forces. For example, Germany currently operates 3,248 tanks, 2,537 armored combat vehicles, 2,058 pieces of artillery, 560 combat aircraft, and 205 attack helicopters. Under the proposed sufficiency rule, Bonn could accommodate an additional 4,752 tanks, 9,463 ACVs, 5,942 artillery pieces, 2,160 aircraft, and 195 attack helicopters. A different threshold or a special arrangement would have to be made to account for Russia's outsized inventory of combat aircraft and helicopters: perhaps that if all holdings are national, a state is subject to only the original Sufficiency Rule.

Provide New, State-Based Counter-Concentration Rules. One such rule might stipulate that no more than *N* percent of a state's forces could be stationed within *M* kilometers of its borders. Earlier assessments suggested that the depth of this buffer zone should be about 60 km, adequate to provide 24 hours' unambiguous warning of an impending attack.[7] Study of the participating states' annual data declarations indicated that there were far too many units within 60 km of frontiers, and that a more-modest buffer zone would have to be acceptable, since the prospects of convincing states to undertake an expensive restationing effort seemed bleak. Our current estimates and map inspections suggest that it might be possible to negotiate a rule to the effect that "no state shall permanently station more than 30 percent of its forces within 20 kilometers of its borders," a measure that would accommodate current force dispositions but that

[7]This figure was derived from old Soviet movement norms for offensive operations, which called for advances in the main attack of 100 km per day. Adjusting this figure downward to account for a lack of funds to support training, maintenance, and readiness, we estimated 60 km to be an optimistic figure of how far anyone operating Soviet-era equipment might advance.

would further retard massing forces to produce an offensive capability. Other useful measures might include mandatory 42 days' notification before temporarily stationing units within 20 kilometers of the border.

Consider Establishing Overlapping Zones. Overlapping zones can assist in ways that single zones cannot. Consider Figure 3.4. Denmark, Germany, Poland, and the Czech Republic constitute the "Blue zone"; Russia, Poland, the Czech Republic, Slovakia, Belarus, and Ukraine make up the "Green zone." Because states belong to more than one zone, there is a greater opportunity to control both the local neighbors who might create trouble and also slightly more remote actors. Each zone might have similar total limits, or a zonal sufficiency rule might be employed, limiting any one state from having more than some agreed-upon percentage of TLE within the zone. In addition, overlapping zones eliminate such problems as one state insisting that it be included in the same zone as its neighbor. Overlapping zones could prevent "a new dividing line in Europe" by establishing agreed-upon limits and constraints across a cluster of states, regardless of their political-military alignments or alliance memberships. Overlapping zones might provide both NATO and Russia with buffer zones in which they could exercise some control over TLE ceilings, military activities, and balance of forces.

Agree to Placing Constraints on Organization of Military Forces. One such measure might be to agree that the majority of states' most-ready forces be deployed in multinational formations. NATO multinational divisions, new multinational structures based on cooperation among NACC and PfP members, the Allied Command, Europe (ACE) Rapid Reaction Corps, the Eurocorps—numerous cooperative structures are already in being under which these formations might be organized. The newly approved NATO CJTFs might be ideal. The 30 signatories might also agree to adopt NATO-like force categories as their norms, maintaining mostly reserve units for their main defense forces and significantly smaller rapid-reaction and immediate-reaction forces for coalition military action. Such a pact would limit states' capability for offensive action against each other while enhancing their ability to conduct coalition operations.

If such high levels of military cooperation still elude the signatories, then perhaps more-traditional organizational constraints would be

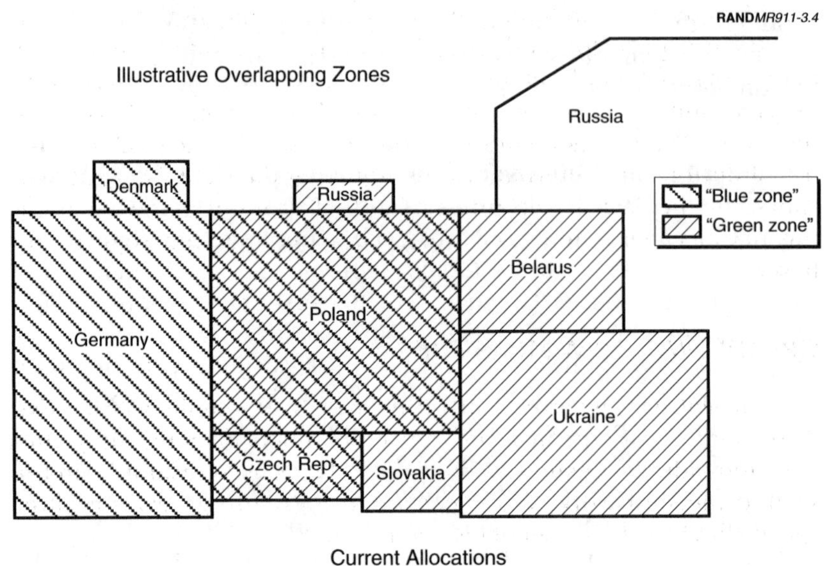

Blue zone:	Tanks	Artillery	ACVs
Denmark	353	553	316
Germany	4166	2705	3446
Poland	1730	1610	2150
Czech Rep.	957	767	1367
Total	7206	5635	7279

Green zone:	Tanks	Artillery	ACVs
Russia[a]	850	426	925
Poland	1730	1610	2150
Czech Rep.	957	767	1367
Slovakia	478	383	683
Belarus	1800	1615	2600
Ukraine	4080	4040	5050
Total	9895	8841	12775

[a]TLE reported in Kaliningrad.

Figure 3.4—The Potential of Overlapping Zones

useful. For example, regulating the amount of artillery available to support maneuver units is one way to limit offensive capability. Rules might be agreed to that would restrict the signatories to no more than one artillery battalion in direct support of any maneuver brigade, or no more than one combat engineer company habitually associated with any maneuver brigade. Individual measures could be negotiated on the mix of combat support units in active and reserve forces.

Our objectives in examining the first group of questions were to assess the prospects for modifying the CFE Treaty and to explore some of the potential pitfalls of those prospects. Analysis of this first group of issues indicates that, while there are obstacles and pitfalls, none need prove fatal to adaptation of the CFE Treaty. The sizable number of options and alternatives for adjusting the treaty's operations should be the basis for optimism about adaptation. But what about the prospects for new measures? The next section reports our findings.

GROUP TWO: NEW MEASURES

Contemplating new arms control measures eventually yielded two distinct efforts. In the first, the project members invented altogether new, individual measures. Using the Balkans as an example, in the second piece of work we concentrated on crafting new measures specifically to satisfy a set of local conditions.

Individual Measures

Karl Pfefferkorn at the University of Virginia noted during the project that it would be difficult to build new stability measures around infrastructure because those facilities are tightly intertwined with a society's economic well-being and therefore are not easily subjected to limitations. He also cautioned that reporting and transparency based on infrastructure were likely to induce an avalanche of worthless reporting. With his caution in mind, the project team determined that there are ways to improve stability through infrastructure measures if the focus of the study moves from the macro level to a more-detailed, specific level. That is, the discussion is not about railroads per se, for example, but about military features and practices of railroads. Thus, in examining possible infrastructure-based stability measures, the project team concentrated on the military features of rail operations, airfields, ports, and road movement that could conceivably lend themselves to new stability measures.

Rail Operations. Rail movement was always a central element of Soviet military operations. It remains the principal transportation mode for movement of troops, equipment, and materiel for Russia and many European forces today. Moving ground forces by rail requires large numbers of flatcars: at least 450 per armored or mechanized brigade. Many railyards intended for marshalling, loading, and shipping ground forces have ramps built by the tracks so that elements of a convoy of heavy or outsized equipment, including armored vehicles, can simply drive onto the flatcars and be tied down.

At least two CSBMs suggest themselves. First, there could be a counter-concentration and notification measure whereby states agree not to assemble more than 1,000 flatcars in any one railyard, and to give 42 days' notice before assembling more than 450 cars to deploy a brigade or other formation for military training or similar activities. A 1,000-/450-car ceiling would not impede normal commerce, would not produce a reporting burden, and would allow for storing cars for maintenance and other activities while restricting the size of a force that could be moved from a single railyard. It would also make enough cars available for military training and exercises. If a state were to undertake military operations, it would first have to disengage flatcars from the commercial sector and concentrate them at or near deploying units, which would be a monitorable, detectable event. States detecting large concentrations of flatcars might query the responsible country and be entitled to clarification within 72 hours or some other short period of time.

Second, states could agree to report and limit the number of loading ramps maintained in their railyards. All depots, railheads, and similar rail facilities with loading ramps might be declared, along with their locations and the number of ramps on each installation. Since Russia and other European countries also use the ramps to load commercial trucks onto trains, it is unlikely that much could be done to limit the use of the ramps. Still, some small percentage of existing ramps might be destroyed—perhaps 10 percent. As an alternative, the number of ramps might be strictly limited at rail facilities that support primarily military garrisons.

Airports. To move a large number of ground forces or a large quantity of supplies by air, most states in Europe must first move the cargo to departure airfields, where dispersion makes them less-vulnerable targets. The same is true for air forces: They leave their home fields and disperse to smaller airfields. Most dispersal airfields are normally used for civil aviation. Some of these airfields have revetments, ready to protect parked military aircraft. Few countries have the assets to maintain large, permanent fuel-storage sites at these airfields; even where the storage exists, the tanks are seldom full because of current fuel constraints. Prior to their military use, the available fuel at these fields will have to be increased and temporary fuel storage established. In addition, airdrop of large military cargo, especially large vehicles, requires special pallets much larger than those used for commercial freight and high-capacity k-loaders to move heavy palletized loads. These k-loaders are also much larger than their commercial counterparts.

Countries might agree to give some amount of notice prior to establishing temporary fuel storage to support military activities. States could agree not to concentrate military pallets or high-capacity k-loaders, or perhaps not to maintain stocks of pallets and k-loaders in the same location.

Ports. Embarking any militarily significant force requires specialized shipping and space for marshalling troops and equipment. Many military force packages would include several roll-on/roll-off (RORO) ships and break-bulk ships. States might agree to notifying other treaty members before assembling more than four ROROs and four break-bulk ships. This approach would not affect commercial shipping, since individual ships would not be reportable—only the combination of ROROs and break-bulk ships would require notice.

Ground Operations. For most European armies, long marches and extended operations depend on having heavy equipment transporters (HETs) to move tanks and other tracked vehicles, as well as tactical fuel trucks to accompany mechanized forces. HETs are typically commercial, but tactical fuelers are organic to the ground forces.

Both HETs and tactical fuelers might be controlled through the use of counter-concentration measures and exclusion zones. For example, states might agree not to maintain tactical fuelers within 200 kilometers of an international border, effectively separating the fuelers from the cutting-edge assault units in any military. States might negotiate a bargain in which they agree that HETs cannot be maintained within the same garrison as tanks and that notice must be given at some agreed-upon interval before HETs can be moved to tank or mechanized units.

Implementation Considerations. The measures sketched out here could be implemented in a way that would not cause worthless and voluminous reporting. As a practical matter, giving notice for the events suggested above would be infrequent, since the thresholds at which notice would be required are set reasonably high. For example, suppose the reporting threshold were 1,000 railcars. Because 1,000 flatcars are rarely sitting in any one railyard, the measure would be useful because it requires reporting only unusual concentrations of cars. Declarations, as of the number of tank-loading ramps, might be made very infrequently. Rather than including them in a state's annual information exchange, they might be reported every two or three years.

Arms Control for Local Problems

The fact that CFE includes special provisions for its flank zones—provisions that limit the buildup of TLE opposite eastern Turkey and Norway—suggests that, under some circumstances, instituting arms control measures tailored to specific, local needs is in order. But what are the prospects for success with local arms control, and how might local pacts be structured? The project team set out to find out, choosing the Balkans to illustrate how to proceed. Note, however, that the deal wrestled out of the former belligerents at Dayton reflects coercive and disarmament-like measures that need not color all local arms control efforts.[8]

The first step involves answering some basic questions: Which local issues and sources of conflict must the arms control and security bargain successfully address? What might such a bargain look like? What does *security* mean in the local context? What will be militarily sufficient to provide security? The notional pact that develops from answering these questions illustrates what kinds of trouble the pact can and cannot address, the need to deal with important issues that contribute to the local dispute but that lie beyond the ability of a security agreement to resolve, and related matters.

---

[8]This effort was conducted concurrently with the Dayton negotiations, so that subsequent developments and agreements were not yet available for consideration.

Managing Sources of Conflict. In very general terms, the objectives of arms control have been to provide force reductions to limit the likelihood of war, to limit the damage if war occurs, and to reduce the burden of defense expenditures.[9] More specifically, recent European efforts, including CFE, have sought to produce stability at lower force levels. In this case, *stability* was understood to mean a greatly reduced prospect of surprise attack across the Inter-German Boundary. The problem in the Balkans is somewhat different, however.

In Bosnia and its neighbor states, mythical and historical episodes of treachery, brutality, and betrayal stretch from 1389 through centuries of Ottoman occupation, the first and second Balkan wars, and both World Wars.[10] The latest spasm of violence and atrocities has provided each of the belligerents with an overlay of personal experience that reinforces their tradition of hatred, fear, and desire for revenge. At issue among the parties is whether they can be treated fairly and justly if they live among a polity dominated by the other factions.

Put another way, each group fears that the rule of law will not protect it from violence at the hands of its neighbors. As a result, each faction seeks a territory of its own to govern. However, much of the land has traditional links to more than one group of people and the emotional attachments are strong—especially among the peasants. As a result, each ethnicity believes that the others have stolen its territory.

The recently concluded Dayton Peace Agreement on Bosnia-Herzegovina offers an opportunity to resolve this dispute by providing a "pretend" federation in which the parties can begin to rebuild their relations—"pretend" at present because at least some of the former belligerents see no incentives for making it succeed, and most of the local actors have doubts about its prospects.[11] Some of the objectives of any local arms control arrangement in such circumstances are to do nothing that would hurt or undermine the

---

[9]Colonel Ralph A. Hallenbeck and Colonel David E. Shaver, eds., *On Disarmament: The Role of Conventional Arms Control in National Security Strategy*, Carlisle, PA: Strategic Studies Institute, U.S. Army War College, 1990.

[10]See Robert D. Kaplan, *Balkan Ghosts*, New York: St. Martin's Press, 1993.

[11]Recent shootings of Muslim police by Croats and retaliatory shots at Croat police by Muslim assailants are symptomatic of the tenuous nature of the bargain.

fledgling confederative relationship and to help it develop into a real federation with the normal scope of interactions among the peoples of Bosnia-Herzegovina.

To deal with these fears and concerns, we gave *stability* in the context of the Balkans a new operational definition that includes components addressing security, civil justice, incentives for cooperation, and confidence and transparency. Figure 3.5 offers a summary.

Security has several necessary components for protecting against internal and external assaults. For Bosnians to be secure, safeguards for their lives and property must be provided and must be backed up by forces adequate to keep the peace and suppress violence. In addition, such forces must be trained and equipped to protect the population (as opposed to merely securing key terrain or protecting themselves).

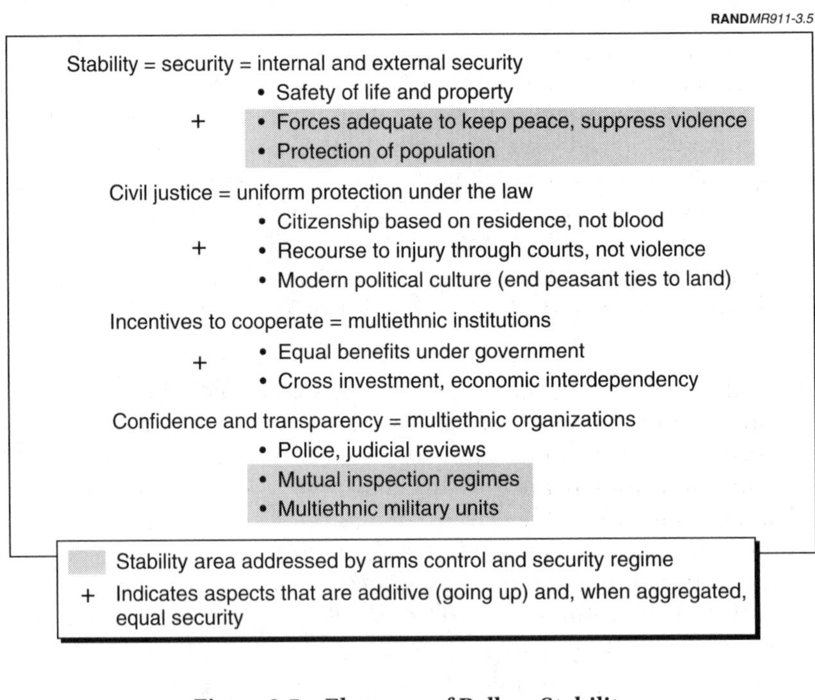

Figure 3.5—Elements of Balkan Stability

Civil justice provides for the rule of law and equal protection under the law. The only way that civil justice can deliver these protections in Bosnia is if the understanding of *citizenship* shifts from the traditional basis of blood roots, which functions to exclude some people, to the modern concept based solely on residence and allegiance, which serves to include more people in the society. Thus, all residents of the region would enjoy equal protections and no one would fall victim to official ill-treatment because they were not blood members of the ruling faction.

An important aspect of *civil justice* is the sense that the courts can ably address injuries, that people can have confidence that if they are wronged the state will provide protection and justice for them and appropriate punishment for the transgressor. It is only when a polity has confidence in its institutions for civil justice that the factions will agree to stand down their ethnic militias and other paramilitary groups and allow the state to exercise control over the use of force.

Incentives to cooperate support civil justice and basic security considerations. Their most basic function is to replace the desire for revenge with the benefits that could arise from cooperation. In the near term, a domestic system that provides multiethnic institutions improves the chance that all members of society will receive equal benefits under the law, because each group has its own representatives in the hierarchy of institutions, presumably safeguarding the group's equities. In the longer term, multiethnic institutions may help integrate society at large and eliminate negative ethnic distinctions.

Finally, *confidence and transparency* in institutions remove suspicions about other factions' motives and provide trust in Bosnian institutions. For example, multiethnic organizations, including the police and appeal and review practices (judicial review, for instance)—ensure that no group is unfairly taken advantage of by the others and that civil justice functions to protect all citizens the same way. *Multiethnic military formations* ensure that no one faction has control of the military instrument and that the forces of the state cannot be turned against the interests of the people at large or against any subgroup. Moreover, these formations are important for the long-term integration of society, since they serve as a venue for discovering cooperation and joint accomplishments. *Mutual in-*

spection regimes give the various factions confidence that they are being dealt with fairly by the other factions. Mutual inspection of military units and their storage facilities assures all groups that no single group is stockpiling or otherwise exploiting what should be mutually held military resources. The inspections also reassure all groups that their conscripts are being fairly treated, no matter what the background of the commander or the ethnic makeup of the unit leadership.

The shaded portions of Figure 3.5, which suggest the areas of stability that an arms control and security regime might address, do not include a majority of the elements that collectively make up "stability." In fact, in the subcategory "civil justice"—arguably the most critical element of stability—the arms control and security regime contributes nothing.[12] That said, an arms control and security regime can contribute to security as understood in Figure 3.5 by sizing and shaping local military forces. In addition, such a pact could help put local forces into a posture in which they can keep peace and protect the population while reducing the likelihood that the military might be used against one ethnicity or another.

As noted, a security regime could contribute little that would directly improve the functioning of civil justice or provide incentives to cooperate. This means that, in addition to a security agreement, the Bosnian factions will need nation-building assistance to develop a modern political culture, raise trust in shared institutions, and foster habits of cooperation for mutual benefit. Given the sources of conflict in the region—emotionally charged territorial disputes, deep-seated distrust of other ethnicities, and a desire for revenge—arms control and security can address only a handful of the issues that underpin the current Balkans conflict. Moreover, it may be that no such security arrangement can proceed until a minimum essential amount of civil justice, confidence, and cooperation among the parties emerges.

---

[12]The Dayton Peace Agreement attempts to provide for building civil justice through the introduction of international police, a commission on human rights, an ombudsman, and an arbitration panel, among other things. See U.S. Department of State, "Summary of the Dayton Peace Agreement on Bosnia-Herzegovina," *Fact Sheet*, November 30, 1995.

Having described some of the things a local arms control bargain might focus on as its working goals—establishing internal and external security, and confidence in official institutions—we now consider the military dimension and the types of arms that might be controlled, limited, or monitored as a step toward achieving these goals.

Military Sufficiency. Central to any arms control deal is the notion of *military significance:* Which weapons constitute the greatest threat and therefore should be controlled? In CFE, the "militarily significant" arms were identified as those contributing to the ability to conduct large-scale offensive operations: tanks, armored personnel carriers, artillery, attack helicopters, and combat aircraft. This treaty-limited equipment gave each group of states the capability for sustained, armored maneuver warfare: both sides' preferred mode of war. By introducing parity between the alliances at lower force levels, CFE reduced the risk of the principal danger, surprise attack, and improved stability.

This subsection examines the same problem set for the Balkans. It examines the local, preferred mode of fighting to learn how it might be hampered by arms control to contain the principal danger: undisciplined attacks on targets of opportunity emboldened by a limited threat of retaliation resulting from the often-significant imbalance of forces that reflects the ethnic compartmentalization of the area. It identifies legitimate requirements for arms in the region, determines which weapons are militarily significant given local military practices, and suggests which types of weapons should be controlled or reduced in a local arms control agreement. From this examination, a sense of *military sufficiency* should emerge: forces adequate for legitimate security needs, but maintained at levels and in postures that will reduce the chance that they will be exploited by any faction.

Virtually all of the six major factions involved in the Bosnian fighting employ infantry forces with varying amounts of augmentation from such supporting arms as artillery and armor (e.g., APCs and tanks). Employment of these supporting weapons is often unconventional: witness news footage of shellings carried out by single howitzers or infantry advancing with a lone tank. There have been no episodes of massed, armored maneuver by any party to the fighting. Most engagements involve small units and relatively cautious movement.

As they seek to reclaim territory to which they feel traditional, historical bonds, forces frequently orient on towns and villages rather than on enemy forces. The targets of these forces have often been unarmed civilians or their property, because the forces seek to drive people of other ethnicities from the land. Rubbling is a common practice to prevent people from returning to their villages.[13] Table 3.2 summarizes the weapon holdings that the various factions use.

Most of the weapons were seized from the Yugoslav National Army (JNA) as fighting spread through the region in 1992. Of the Bosnian internal factions represented by the top three entries in the table, the Bosnian Serbs succeeded in capturing the largest share; the Muslim faction seized relatively little. The last three rows in the table reflect the holdings of Bosnia's neighbors involved in the fighting.

Two features of Bosnian warfare suggest that many of the weapons listed in the figure are militarily significant in the Balkans and, hence, good candidates to become treaty-limited equipment in any new arms control pact.[14]

First, terror strikes have been widely practiced. In these attacks, a single mortar bomb or recoilless rifle shot is delivered into a crowded civilian area: typically, a marketplace, bus stop, or cafe. Often small-caliber, the weapons employed would not be controlled as TLE under the CFE Treaty. In the Balkans, however, they are *politically significant* because they demonstrate to the people that the government cannot protect them. Thus, when the people leave the region to escape the danger, the attack takes on *military significance* as well, because it has helped the attacker achieve its goal of securing disputed territory.

Second, air defense and anti-tank guns, long obsolete in their original roles, have been deployed as direct-support weapons for ground forces. They have been used against fixed targets, including homes

---

[13]*Rubbling* is deliberate destruction of buildings, usually to create an obstacle or barrier.

[14]The arms control negotiations in Bonn considered only CFE-defined TLE.

Table 3.2

Comparative Balkan Arsenals

Faction	MBT	APC-AIFV	ARTY	MRL	Mortar	ATGW	AD-AT guns	Helo	AC	Troops
Muslim Bosnians	31	35	100	2	200–300	100+	some	5	3	92,000
Croat HVO	75–100	80	200	30	300			6		50,000
Bosnian Serbs	370–460	295–400	730–1200	76–106	900		some	24–45	50+	75,000
Croatia	181–560	255–320	949	21	761–1000	180–4000+	700+	5	28	99,600
Serbian Krajina	240–250	100	200	14	some	some	30+	9–12	16	40,000
Yugoslavia	639	629	1499	72	2266	135	1864–3500	110	1714	126,500

SOURCE: Compiled from IISS, *Military Balance 1995/96*, London: Oxford University Press, 1995, and the U.S. Naval Institute database, *Periscope*, August 1995.

NOTE: MBT = main battle tank; APC-AIFV = armored personnel carrier–armored infantry fighting vehicle; ARTY = artillery; MRL = multiple rocket launcher; ATGW = anti-tank guided weapon; AD-AT = air defense–anti-tank; Helo = helicopter; AC = aircraft; and HVO = defense council.

RAND *MR911-T3.2*

and barns, and against enemy forces. There is at least anecdotal evidence that these weapons have been the principal instruments of death in a number of Croatian mass murders: Huge numbers of spent shell casings have been found at sites of suspected mass graves.[15] Anti-tank guided weapons (ATGW) perform the same function. Although more expensive than earlier anti-tank weapons, the ATGW are held in such abundant numbers that they too should be controlled because of their potential for damage.

Based on this cursory summary of weapons employment and Balkans combat practices, an expanded set of TLE suggests itself. In addition to CFE TLE definitions that provide control of tanks, artillery, armored personnel and infantry fighting vehicles, attack helicopters, and combat aircraft, several new categories would be useful. First, the definition of *artillery* should be expanded to capture howitzers, guns, and mortars to as small a caliber as 81 millimeters. ATGW might also be a category. A new category for air defense and anti-tank guns should be created to control air defense guns to as small as 23mm and anti-tank guns to 76mm.

Force Size. Of course, as with any other state, Bosnia has legitimate needs for armed forces. The next issue, therefore, is to determine the appropriate size of these forces so that the regional balance will remain undisturbed or so that an imbalance will not touch off the security dilemma among the neighbors—while providing forces adequate to secure the population and its property from internal and external foes. Once the basic dimensions of the force are established, TLE distribution, force organization, and related matters can be determined.

Traditional approaches to defensive force sizing generally focus on force-to-space ratios.[16] In these approaches, force designers consider how much frontage they have to defend or the numbers of avenues of approach into their territory that the forces must control.

---

[15]For sample casualty reports, see "The War Against Croatia (1991–1995)" on the World Wide Web at http://www.tel.fer.hr/hrvatska/WAR/WAR.html/ (downloaded January 10, 1996).

[16]For a good review, see Wayne P. Hughes, ed., *Military Modeling*, 3d edition, Washington DC: Military Operations Research Society, 1986, especially Chapters 4 and 9; Joshua M. Epstein, *Conventional Force Reductions*, Washington DC: The Brookings Institution, 1990, Chapter 5.

Force planners also consider the military prowess of their adversaries and the force size necessary to ensure favorable force ratios along all invasion routes.

The first step in determining an appropriate force size for Bosnia is doing a comparative analysis of the neighboring states. Such an analysis, summarized in Table 3.3, compares the relationship between the neighbor countries' population, size of territory in square kilometers, and the size of its army per kilometer of border length.

The table indicates that regional armies vary greatly in size from Macedonia's 10,400-man force to Romania's 128,800-strong establishment. Moreover, both Croatia and Yugoslavia[17] have sizable forces and both have been involved in the Bosnian War. If Bosnia were to build its army along the regional norms from the "Average" row in Table 3.3, based on its size and population, a force ranging in size from 31,500 to 48,600 persons would emerge. However, given the recent history of invasion and indirect involvement of its neighbors plus the requirement for internal security, it would be more prudent if Croatia, Bosnia, and Yugoslavia "built down" (a process in which Bosnia would build up its forces while the others would reduce theirs to the same or similar levels) their forces to parity at a somewhat higher level. If an army of 70,000 were the build-down target, all three states would make reductions of 20,000 or more. Such a size would not change the proportional relationship with other states in the area: All three would still have larger standing forces than Albania, Bulgaria, Hungary, and Macedonia, and all three would remain smaller than Romania's forces.

In the process of establishing a built-down target level, it is important to remember that Bosnia not only needs parity with its external antagonists but must also have internal, factional safeguards: Whereas Bosnian Croats and Serbs can count on reliable outside assistance from their kinsmen in neighboring states, Muslims have few outside friends. Since Muslims constitute approximately 44 percent of the Bosnian population and occupy approximately 40 percent of its territory, they should make up around 42 percent of Bosnia's standing ground forces. Serbs, at 31 percent of the population and

[17] *Yugoslavia* refers to the rump state of Serbia and Montenegro.

Table 3.3

Comparison of Balkan States and Armies

State	Population	Territory (sq km)	Border Length (km)	Army Personnel	Soldiers per km of Border	km^2 per Soldier	Population per Soldier
Albania	3,535,000	27,400	720	60,000	.01	.45	59
Bulgaria	8,411,000	110,550	1808	51,600	.04	2.14	163
Croatia	4,785,000	56,538	2028	99,600	.02	.57	48
Hungary	10,206,000	92,340	1989	53,700	.04	1.72	190
Macedonia	2,228,000	24,856	748	10,400	.07	2.39	214
Romania	22,805,000	230,340	2508	128,800	.02	1.79	177
Yugoslavia	10,821,000	102,136	2246	90,000	.02	1.13	120
Average				70,586	.03	1.46	139

SOURCE: Compiled from IISS, *Military Balance 1995/96*, London: Oxford University Press, 1995, and the *Central Intelligence Agency World Fact Book 1995*, Washington, DC: U.S. Government Printing Office, 1995.

RAND*MR911-T3.3*

controlling around 49 percent of the land, should constitute about 40 percent of the army. Croats would provide the remainder. Based upon such a formula, Bosnian Muslim forces would number 29,400; Serbs, 28,000; and Croats, 12,600.

Distributing TLE. Distributing TLE to the three factions within the Bosnia-Herzegovina army should take into account the following three points.

First, since all factions are descended from the former state of Yugoslavia, they should have equal claims on the equipment of that state's military, the JNA. In other words, all parties should have equal access to former JNA equipment in order to establish military equilibrium in the area. The fact that some factions seized larger stocks than others during the civil war should not prejudice each group's claim for a share of the armaments commensurate with legitimate security requirements.

Second, the distribution of TLE should be congruent with force structure. A faction infantry brigade, for example, should have TLE appropriate to such a formation and not have inflated holdings of tanks or other weapons.

Third, the force structure adopted by the army of Bosnia-Herzegovina should be appropriate for the terrain and local military capabilities. For example, none of the factions would be expected to be discovered building a combined-arms mobile strike force of corps strength, since none of the military actors in the region are competent to command and control such a formation and the formation is not appropriate for Balkan terrain.

The foregoing points suggest that each of the factions should build a brigade-based force of principally infantry units. Figure 3.6 outlines the possible organization, which results in a brigade of 3,445 troops and a TLE density that is on a par with that of many European forces.

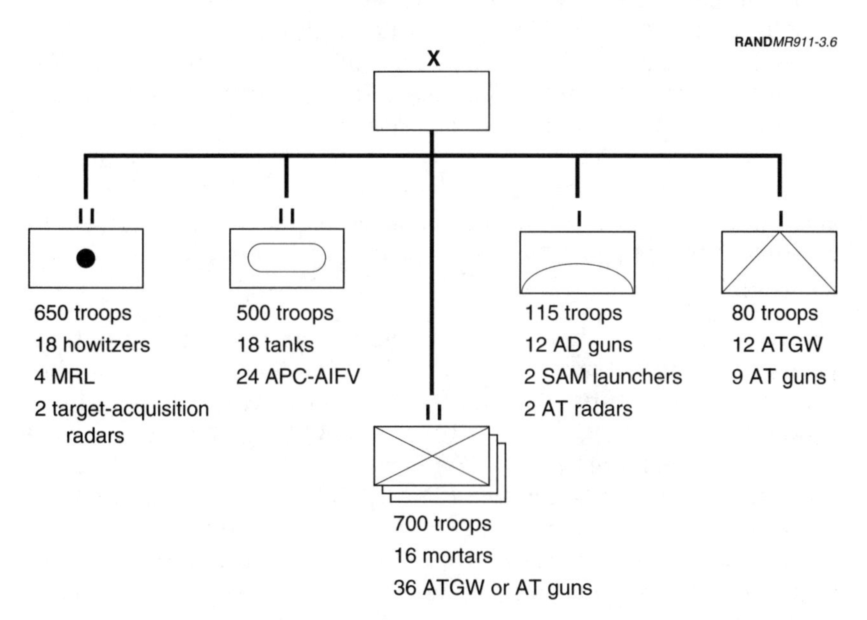

Figure 3.6—Notional Bosnian Infantry Brigade

The organization provides proportionate artillery, air defense, mechanized, and anti-tank forces for a predominantly infantry brigade containing three infantry battalions. With its armor, anti-armor, and artillery, this force design offers limited offensive capabilities but sound self-protection. To make it capable of independent combat operations, the design could be augmented with organic supply and maintenance.

If this force structure were adopted and JNA TLE redistributed across it, the Muslim portion of the force would produce about 8-1/2 brigades; the Serbs, 8 brigades; and the Croats, about 3-2/3 brigades. The aggregate force structure—units of all three factions—would deploy roughly 19 brigades with TLE and related arms totaling about 342 howitzers or field guns, 76 multiple rocket launchers, 228 air defense guns, 38 surface-to-air missiles, 342 tanks, 456 APCs or armored infantry combat vehicles, 912 anti-tank guided weapons, 171 anti-tank guns, and 304 mortars of various calibers. Such a force, roughly equivalent to 6 divisions, would be almost as large as the U.S. force deployed to Vietnam near the height of U.S. involvement there. The forces deployed in Vietnam were arrayed across approximately 211,480 square kilometers; by comparison, Bosnia's territory constitutes only 51,233 square kilometers. Although Bosnia's mountains offer different obstacles than the rice paddies and highlands of Vietnam, six infantry divisions should be perfectly adequate to defend Bosnian territory against outside incursions and to provide internal security.

Of course, such a scheme would produce an "overhang" in armaments: equipment available in Bosnia, Croatia, and Yugoslavia but not allocated to units. In accordance with the 19-brigade force structure, up to 1,698 tanks, 1,108 APCs, 3,806 artillery pieces, 169 multiple rocket launchers, 4,462 mortars, 3,323 ATGWs, and 4,002 air defense and anti-tank guns would be available from JNA totals. Even after distributing TLE to Croatia and Yugoslavia, some overhang remains. Nor does the brigade-based force structure deal explicitly with attack helicopters and combat aircraft—major force multipliers in the region. Storage arrangements could be made for the excess TLE, which would leave all parties confident that no single group would be able to steal valuable arms and leave the others at a disadvantage. Aircraft and helicopter holdings might be centralized under a multiethnic national control.

An Arms Control and Security Regime. Having briefly explored the origins of conflict and the military arsenals that support fighting in the area, it remains to be seen what kind of arms control and security regime might be useful. The working definition for *stability* proposed earlier in this subsection (presumably the objective of any arms control bargain) emphasized security, civil justice, cooperation, and confidence and transparency. That discussion also acknowledged the limited ability of an arms control pact to satisfy fully the requirements for stability. However, the Dayton Agreement provides a framework for managing issues of civil justice and cooperation, so the following proposed arms control and security arrangement builds on the Dayton Accords while also providing militarily significant security, confidence, and transparency measures.

The proposed accord includes a near-term (first five years) and a long-term (perhaps 10 years) phase. During the near term, the emphasis is on building down forces and establishing confidence and security; the long-term objective is reintegration of Bosnian armed forces on a multiethnic, cooperative basis. During the first five years, the Bosnian factions would control their share of the TLE, consisting of tanks, APCs-AIFVs, combat aircraft, and attack helicopters, as they are understood under the CFE Treaty. In addition, the Bosnian arms accord would also control artillery and mortars to 81mm, anti-tank guided weapons defined by a protocol of existing types,[18] air defense guns to 23mm, and anti-tank guns to 76mm. Former JNA TLE would be distributed proportionately among the Bosnian factions, according to the territory apportioned to them under the Dayton Agreement and according to their respective population size as suggested earlier in this chapter. In the event that adequate TLE cannot be found to equip all factions appropriately, NATO or other interested parties might contribute the arms necessary to bring any faction up to the level of holdings to which it is entitled.

The actual level of active forces and TLE would be negotiated, but the objective should be to reach an agreed-upon level of forces and a common force structure based on common units such as the brigades suggested earlier. The approach would be to build down to

[18]Such a protocol would seek to capture all the weapons in use in the area: Milan, Hot, Sagger, etc., ATGMs, and would not be confined simply to weapons of a given caliber or greater.

force levels adequate to provide security while reducing the number of arms and units in the area. Units would be organized to include reasonable amounts of TLE, again along the lines suggested in the notional brigade model of Figure 3.6. The aggregate forces from all Bosnian factions would be adequate to defend the state.

The TLE beyond that needed in each faction's active forces would be held in supervised storage sites under *dual-key control*: The sites would be situated on the territory of their respective factions and would be operated to allow normal maintenance and training activities. Each site would be manned and secured by its own faction. A second perimeter, manned and secured by another faction, would be drawn around the site to ensure that TLE designated for supervised storage was not deployed to provide the other faction an advantage in active forces. Combat aircraft and attack helicopters would be held as the state strategic reserve under multiethnic, centralized control.

During the first five years of the accord, Bosnia would promote the professionalism of all its armed forces and work toward greater cooperation among the factions. A mutual inspection and verification measure would be in place, involving all of the factions in a series of visits to each other's units to verify that the other parties had only the active forces to which they were entitled under the agreement, and that those forces were adhering to the agreed-upon force structure and TLE constraints.

Faction forces would exchange observers and monitors and participate in joint exercises under the auspices of the Organization for Security and Cooperation in Europe (OSCE), NATO, or some other body. The member states of the OSCE and other interested parties would make available civil-military–relations programs and military professionalism training. In addition, the factions would undertake a joint program to locate, render harmless, and destroy the thousands of mines that have been sown across the countryside. Through training, joint exercises, and real (i.e., goodwill-producing as opposed to committee-forming) cooperative programs such as the counter-mine effort, mutual respect—and eventually trust—may emerge and gradually replace factionalism and confrontation within the Bosnian federated military forces. This is a tall order, one not achieved during the near half-century of coexistence within the Yugoslav state. But perhaps after the most recent bloodletting, and aided by the measures suggested here, more-rational minds will prevail.

After the first five years of the agreement, focus would turn toward reintegration of Bosnian forces and the creation of multiethnic military districts and brigade groups that would organize once-separate factions within a single unit. Force levels and TLE apportionments would remain in place. Bosnia would move toward a combined command council for control of its armed forces, ensuring that each faction was represented and that authority and key positions were shared equally. As a sense of civil justice, civil society, and modern political culture matures through other initiatives outside this arms control accord (e.g., through the OSCE, EU, NATO and other activities), the Bosnians may eventually have enough confidence in their armed forces that reorganization of the armed forces along functional, rather than multiethnic, lines will be possible.

Observations About Local Arms Control. An arms control program like that suggested here could make a significant contribution to local stability. Under ideal circumstances and when implemented together with programs aimed at restoring civil justice and a normal society, arms control could help transform Bosnia from its current, miserable conditions into a real multiethnic federation with sound prospects for the future. Such an outcome requires that all parties build on the Dayton Agreement and its "pretend" federation, and enlist the aid of regional organizations to help promote civil justice and develop incentives for broader cooperation. The OSCE and others could, for example, provide human-rights monitors, help establish impartial police and judicial oversight, and take other actions to ensure that Bosnia's civil institutions function justly for all citizens, regardless of their factional membership or ethnic identity. Joint commercial ventures that involve multiple ethnic groups and thus teach cooperation for mutual gain would also be useful.

Such an ambitious course of action requires not only the commitment of the factions to return to peaceful relations but also the will of the Europeans and the United States, manifested through the regional international organizations, to support civil justice. It remains to be seen whether such will exists. If it does not and arms control must forge ahead without additional incentives for cooperation or tools to promote the impartial rule of law, the contribution to local security that arms control can make would necessarily be more modest. Nonetheless, if an arms control and security regime can reform and professionalize military practices so that the local armed forces are not directed against unarmed civilians and if forces can be reduced to levels commensurate with legitimate internal and international security requirements, the effort to craft the agreement will have been worthwhile.

Negotiating an Arms Accord. How to negotiate such an arms accord? The deep mutual suspicions among Bosnia's factions suggest that negotiations can only begin with outside support: with the credible good offices of a major power. Having sound sponsors such as the United States to underwrite the negotiations would be crucial, as with the Dayton Agreement. If the United States is to preserve the perception of its impartiality and thus its suitability as a sponsor for local arms control negotiations, the U.S. government will need to balance carefully its role and actions as the leader of the Implementation Force (IFOR) and Stability Forces (SFOR) against future requirements for U.S. leadership in brokering an arms pact. That said, some circumstances will produce conditions beyond Washington's control. For example, if SFOR's mission proceeds smoothly, the United States should remain a credible sponsor for Bosnian arms control. But if the situation degenerates into violence, there will be little immediate need for arms control.

How to proceed? Assuming that the United States wants to promote local arms control in the Balkans, the U.S. government might proceed as follows:

- Since Germany has already provided a venue (Bonn), there is no need to return to Dayton. The United States can, however, offer the same sort of diplomatic catalyst that propelled all parties toward the Dayton Agreement.
- Offer tagging technologies that make possible remote monitoring and tracking, and national technical means to support all parties during the reallocation of JNA TLE and movement of the nonactive TLE into supervised storage sites. Tagging, developed principally for nuclear arms control under the Cooperative Threat Reduction Program for demilitarization and safe storage of Soviet nuclear weapons, could be used to allay fears that TLE

might be stolen or diverted during a redistribution plan. If the factions conclude that TLE may disappear from accountability during any reallocation scheme, they will be reluctant to participate—especially if their faction is called upon to give up an advantage in certain arms. Tagging arms shipments and using national technical means (e.g., intelligence satellites) to track arms transfers might make it easier for some parties to agree to TLE reallocation measures.

- Offer to "round out" factional TLE allocations if JNA stocks are inadequate to equip all parties fully. Since factions that have an advantage in armaments may be reluctant to relinquish it, the United States should be prepared to supply equivalent arms to the less-well-equipped parties. This is not to advocate a buildup policy, but to acknowledge that entities with a clear advantage (e.g., the Bosnian Serbs with artillery) may not negotiate to parity and that it might be necessary to provide similar arms to the others in order to reach an agreement.
- Offer Open Skies aircraft to support monitoring and observation of factional military exercises and movements. Since Open Skies has not been ratified and entered into force, U.S. and European Open Skies aircraft might be made available to monitor Bosnian military activities. All the factions could be provided with mission reports, thus reducing ambiguity about suspect military activities. The aircraft could fly routine monitoring missions above known and suspected garrisons, and could be deployed to cover exercises and troop movements.
- Assist all parties in devising mutual inspection and verification measures. The United States has negotiated several accords that include inspection and verification measures (CSBM Stockholm and Vienna agreements, CFE, START) and could use this experience to help the Bosnians negotiate something similar.
- Volunteer military-to-military training teams to help professionalize Bosnian forces of all factions. The United States has used military-to-military contacts successfully to establish rapport with and to assist former Warsaw Pact forces in learning how to operate as the military instrument of a democratic state (International Defense and Development and International Military Education and Training programs, for instance). The United

States could likewise teach the Bosnian factional forces to respect the rule of law and the Geneva Conventions, and to help them learn professional standards of military conduct.

- Provide seats in U.S. service colleges to promote officer professional development and an understanding of the military's role in society. The Marshall Center, the war colleges, and the U.S. Army Command and General Staff School each offer officers of various grades insights into the role of the military in a democratic society. Promote a similar program through friendly and allied European states. A few Bosnian officers sent to these institutions could become the "seed corn" for developing a modern, Western civil-military relationship and perhaps even a modern political culture.
- Promote initiatives by European states to contribute to the rebuilding of civil justice and a healthy society. Advocate steps that would lead to developing political culture and civil society in Bosnia.

This section has recounted the project team's examination of the prospects for incorporating new measures into CFE, including new categories of TLE and subregional agreements to address local security issues. The next section summarizes our attempts to explore new avenues through modeling.

GROUP THREE: THE MODELING EFFORT AND THREE SCENARIOS

After having established that the CFE Treaty in its present form did not focus on the most compelling questions in contemporary European security, the project team sought to explore ways in which the treaty could be modified to address current European security concerns. What should the new arms pact look like?

To answer this question, the project team:

- examined alternative versions of potential new CFE treaties
- tested each alternative against different scenarios

- compared the relative strength and weaknesses of the different treaty versions
- drew conclusions from its comparisons.

The project posited three alternative future CFE treaties and tested them with scenarios that illustrate some of the signatories' fears: a Russian attack on Poland and conflicts between Hungary and Romania and between Greece and Turkey.

In Alternative One, CFE's legal basis was altered from the bloc-to-bloc framework, but today's allocation of treaty-limited equipment to individual states did not change.

The first CFE adaptation seeks to preserve a "good deal"—a mechanism that allows the participants to keep tabs on Russia. In this alternative, many of the signatories decide that the original treaty was useful in providing stability at lower force levels, and especially for giving the other 29 states an opportunity to monitor Russian military activities closely. Their goal, therefore, is to preserve as much of the original treaty as possible, making only minor concessions to demonstrate goodwill.

As a result, Alternative One maintains current TLE levels. The groups-of-states structure has been replaced. Each of the 30 participating states has a national entitlement that corresponds to its current TLE allocation, summarized in Table 3.4. The number of inspections each state is obligated to accept (its passive inspection quota) is based on the number of objects of verification it has declared: 15 percent of its OOVs for declared sites and 23 percent for challenge inspections during the "residual period."[19] Active inspection quotas are determined by aggregating the number of OOVs in the treaty area and dividing by 30, providing each state with approximately 35 active inspections.[20]

---

[19]According to the treaty, the "residual period" began March 16, 1996, and extends so long as the treaty remains in force.

[20]Since the On-Site Inspection Agency tracks only data on Eastern states, we calculated active inspections based on the number of OOVs associated with those states.

Table 3.4

National TLE Entitlements

	State	Tanks	Artillery	ACVs	Aircraft	Helicopters
West	Belgium	334	320	1,099	232	46
	Canada	77	38	277	90	13
	Denmark	353	553	316	106	12
	France	1,306	1,292	3,820	800	352
	Germany	4,166	2,705	3,446	900	306
	Greece	1,735	1,878	2,534	650	18
	Italy	1,348	1,955	3,339	650	142
	Netherlands	743	607	1,080	230	69
	Norway	170	527	225	100	0
	Portugal	300	450	430	160	26
	Spain	794	1,310	1,588	310	71
	Turkey	2,795	3,523	3,120	750	43
	U.K.	1,015	636	3,176	900	384
	U.S.	4,006	2,492	5,372	784	384
	Unallocated	858	1,714	178	138	0
East	Armenia	220	285	220	100	50
	Azerbaijan	220	285	220	100	50
	Belarus	1,800	1,615	2,600	260	80
	Bulgaria	1,475	1,750	2,000	235	67
	Czech Rep.	957	767	1,367	230	50
	Georgia	220	285	220	100	50
	Hungary	835	840	1,700	180	108
	Kazakhstan	0	0	0	0	0
	Moldova	210	250	210	50	50
	Poland	1,730	1,610	2,150	460	130
	Romania	1,375	1,475	2,100	430	120
	Russia	6,400	6,415	11,480	3,450	890
	Slovakia	478	383	683	115	25
	Ukraine	4,080	4,040	5,050	1,090	330

SOURCE: Adapted from Richard A. Falkenrath, *Shaping Europe's Military Order*, Cambridge, MA: MIT Press, 1995, Tables 3.3 and 6.3. Copyright © 1995 by the Center for Science and International Affairs, John F. Kennedy School of Government, Harvard University, Cambridge, MA. Reprinted by permission.

RANDMR911-T3.4

In Alternative Two, national TLE entitlements were cut by 50 percent and a new group of CSBMs was introduced. This alternative uses CFE as a hedge against the return of the "bad old days"—against the rise of a regional hegemon, perhaps Russia, Germany, Turkey, or others, or against the prospect of intra-European instability and arms racing. In this concept, the adapted treaty includes 50-percent reductions in TLE entitlements, additional counter-concentration measures to prevent any signatory from massing militarily significant offensive forces, more events requiring advance notice and observers to improve transparency, and subzonal measures to ensure military equilibrium in areas of heightened local tensions.

Alternative Three was modest and falls between the first two alternatives: TLE was reduced 10–15 percent, and another CSBM package was introduced so that NATO enlargement is not hampered by the ceilings on the TLE each group of states is entitled to maintain. NATO can accommodate all the TLE of new members. Finally, this alternative seeks to reassure countries that do not wish to join the Atlantic Alliance by giving them additional TLE allocations.

The Test Scenarios

Each CFE alternative meant that a different configuration of forces would be available for the fighting in each scenario. Using a combination of START, a spreadsheet-based Lanchestrian model to simulate the combat ensuing from the scenarios, and some wargaming to account for some of the effects of CSBMs, we were able to assess the relative merits of each treaty alternative and to design a proposed new treaty emphasizing the strengths and downplaying the weaknesses uncovered in the scenarios. The modeling effort was modest, used principally as an aid to thinking, but its results suggest in broad terms what various forms of adaptation might deliver.

Russia and Poland. In the first scenario, Russia (with Belarus complicity) conducted a limited-objective attack into Poland to seize land routes to Kaliningrad. The operation was of short duration, with a cease-fire ultimately brokered. The attack was conducted without a buildup of forces, because Moscow feared that any significant movement of forces or aircraft would tip its hand. Poland received little tactical warning as Russia and Belarus conducted standing-start offensive operations into the Warsaw Military District. Russia attacked out of Kaliningrad toward its objectives, the towns Goldap and Suwalki, while Belarus attacked toward Augustow, securing them and asking for a cease-fire before Poland could reinforce from its Pomeranian Military District or seek help from NATO or western Europe.

Hungarian-Romanian Border Clashing. The two governments traded charges and countercharges of provocations surrounding a Hungarian-minority incident in Romania, in which demonstrators were arrested for displaying Hungarian flags. For reasons that remain obscure, border guards began trading small-arms fire near the Hungarian city Nyírbatór. Romanian internal security forces moved to Carei, a small Romanian town opposite Nyírbatór. Neither force attempted a cross-border operation; rather, they engaged in static firefighting and observing, with each exchange escalating slightly from small-arms fire to light mortar fire, and ultimately to air strikes and artillery duels. The crisis culminated in a brief ground offensive and counterattack. Budapest and Bucharest submitted to outside arbitration to settle the dispute.

Greco-Turkish War. The long-festering grievances between Greece and Turkey unfolded in a new venue: Thrace. Frustrated over their inability to resolve the Cyprus problem and fueled by their more recent clashes in the Aegean, both parties were spoiling for a confrontation in Thrace, where they could bring their land forces to bear. It remained uncertain which side initiated actual hostilities, but each side appeared determined to humiliate the other. After several weeks of combat, whose outcome will be determined by the simulation, both sides remain resolute and have refused international intervention and crisis-management efforts.

Outcomes

Running the three simulated crises under the constraints imposed by each new CFE Treaty alternative, the modeling exercise suggested several limitations and potential contributions from an adapted treaty. Only three excursions were ultimately run. The outcomes reported below are, therefore, highly tentative.

Alternative One (No Real Change from Current TLE Levels). This vision of an adapted CFE Treaty was not particularly useful in controlling or limiting any of the crises posited in our explorations. The zonal structure, unchanged from the current treaty, provided no counter-concentration measures to limit the offensive capabilities in Kaliningrad or Thrace. The forces that initiated the Hungarian-Romanian border clashes were not constrained by the treaty in any way; they were paramilitary and security forces fighting with weapons that did not qualify as TLE. As the Hungarian-Romanian crisis escalated, TLE did enter the fight, but at densities well below the levels "managed" by this version of CFE.

Alternative Two (50-Percent TLE Reductions, Aggressive CSBM Package). Deep TLE reductions could make some military operations untenable. Few military planners would undertake the limited-objective attack into Poland with such force levels. The Greco-Turkish fight would also be more problematic. However, border clashes, like that posited between Romania and Hungary, remain unaffected. Such deep reductions would limit the military capability of the entire region, having, perhaps, the effect of dulling the Eastern and Central European sense of urgency about gaining NATO membership and blunting Moscow's reaction when the alliance enlarges.

Alternative Three (10-Percent TLE Reductions, Moderate CSBM Package). Moderation does not pay. Small TLE reductions did not change the basic dynamics of the scenarios used here. Reductions of 10 percent did not make the limited-objective attack impossible, nor did they constrain border clashes or Thracian adventures.

General Observations from Modeling

Attack helicopters make a huge difference. They provide operational flexibility, mobility, and firepower. In each of the scenarios, the addition of helicopters (and, to a lesser degree, combat aircraft) produced a major advantage, even when other TLE was kept at low levels. A state enjoying a relatively large inventory of attack helicopters and combat aircraft confronting an adversary with a relatively small force is thus very destabilizing, because it creates incentives for preemption. This suggests that new constraints on these two treaty-limited items deserve study.

Tough CSBMs also make a difference. Measures that prevent or delay force concentrations or that constrain surprise become more effective as TLE levels drop. The reason is obvious: If a state has limited military forces and is not confident that it can concentrate them secretly to mount a successful attack, it has less incentive to attack. Monitored-storage (e.g., TLE restricted to a motorpark and training to inside the post, monitored by foreign observers or sensors—constraints removable only with prior notice) monitors and observers at more events and facilities, and more events requiring prior notice become more useful with major TLE reductions, because, at low TLE levels, even a small effect from CSBMs will have a major impact on the available forces, slowing their assembly, compromising the secrecy of their operations, and generally rendering them less effective.

Relevance to CFE Adaptation

CFE adaptation will combine military and political factors, so these modeling results are hardly the final word. But they do plainly suggest military directions to pursue.

Further TLE Reductions. To make a military difference, reductions would have to be large, such as the 50-percent reductions modeled above. Lesser reductions might be politically useful, but they run the risk of appearing to matter when they do not. Deep reductions bring with them questions about how to pay for the destruction of additional TLE, an issue for Russia and some Eastern and Central European states that object to the expense of reductions. Deep reductions also would bear on NATO expansion by making it more palatable to those who oppose it. It would also remove the issue of rapid reinforcement of new alliance territory, since Russia, Ukraine, Belarus, and other states within the treaty area would have little offensive capability. With such deep reductions, the entire treaty area would become a collection of states with sound defensive capabilities and little or no offensive, power-projection capability. NATO enlargement under these circumstances would have far fewer military consequences, since the alliance would be garnering far fewer military resources.

CSBM Packages. CSBMs in the Vienna Document of 1994 have been viewed with caution, even suspicion, in part because they have been politically rather than legally binding.[21] But they can be stabilizing, especially when linked with deep TLE reductions. New storage measures, counter-concentration measures, monitoring and reporting features, and more inspections and verifiable events all add greater transparency to the region and make it difficult to amass a force adequate for offensive operations and, at the same time, to maintain surprise. Thus, the CFE Treaty would become a significant source of reassurance and stability in its own right, reducing the consequences of being inducted into NATO or excluded from membership.

A new class of CSBMs—perhaps called *defensive force structures*—might be studied. All treaty states with forces in an agreed-upon subzone or zones would assume a territorial defense posture for all their forces except a small immediate-reaction force. Combat support and combat service support formations would be based in geographically fixed depots. The air defense network would be fixed. Thus, the states in the defensive-posture zones (or perhaps the entire treaty area) would be able to defend themselves but not to project much force. Moreover, to act against an external threat, the members would have to cooperate to mount an effective force. This posture might ease Russian fears about NATO enlargement.

The effect on NATO would be to remove any urgent need for defense modernization. If Russia and the other treaty signatories are all locked in defensive postures that can be verified on a regular basis, there can be little premium associated with additional defense spending. The limited amount of convertible currency available in Eastern and Central Europe could be devoted to economic revitalization and the improvement of public infrastructure. In addition, since no one state would have a major power-projection capability, such a posture would make Russia a real—as opposed to merely a rhetorical—NATO partner in dealing with threats at and beyond Europe's frontier.

Zones. Overlapping zones and additional subzones, created to regulate the behavior of neighbors or small clusters of states, could be useful. New zones might simply be added to the existing zonal

---

[21] *Politically binding agreements* are promises between states without the force of law. *Legally binding agreements*—treaties—must be ratified by a state's parliamentary body as well as embraced by the chief executive, and carry the weight of law.

structure; there is no reason to remove the current zones, which might prove useful in some future circumstances. Any new subzone structure need not stretch across the entire treaty area; new zones might be negotiated just among the states concerned. These tools might have special bearing on the flanks, engaging Russia with Norway and Turkey in specialized arrangements.

Inspections and Verification. The number of inspection quotas provided under the current treaty will be inadequate over the long term. Current practice indicates that many states will want to inspect Russia, quickly exhausting Moscow's passive inspection liability. Local tensions could have the same effect among neighbor states. Negotiating significantly larger inspection quotas will probably prove impossible, because no state will want either the expense of the additional inspections or the level of scrutiny necessary to satisfy all the other states. A new approach may be warranted. To expand and elaborate upon the functions of the current Verification Coordinating Commission, the members might negotiate a treaty-verification office as a subordinate element of the Joint Consultative Group (JCG). The verification office would share all inspection reports with all parties and would relay concerns from all to any group of states about to conduct an inspection, thereby making the most of each inspection. A treaty-verification office would improve transparency and give Russia a better look at an enlarged NATO—and vice versa.

New Types of Treaty-Limited Equipment. The original categories of TLE were predicated on their military significance for offensive operations in the Central Region. The Bosnian experience testifies to the lethality in Europe of even obsolete anti-aircraft artillery systems and other crew-served weapons used in a ground support role. The current 2-meter rule on spaces subject to challenge inspection, which allows inspectors access to any space at least 2 meters square, should be adequate to control for at least some of these arms. Negotiating one or two new categories of ground support direct-fire weapons with an accompanying protocol on existing types could reduce the level of violence when local conflicts escalate to open fighting.

In this chapter, we have looked carefully at three groups of issues, including the questions surrounding adaptation of the CFE Treaty, the potential for additional measures, and the prospects for fitting CFE effectively to contemporary European security concerns. We conclude that, while the treaty can certainly be adapted to make it more suitable for present security considerations, CFE is unlikely to be as salient to Europe's current worries as it was during the last years of the Cold War. Having picked through the details of the treaty, we return in the next chapter to a more strategic perspective.

Chapter Four

THE BIGGER PICTURE

CONCLUSIONS

At this point, unless the talks derail because of the NATO-Russia dispute over alliance enlargement, there is every reason to believe that the CFE Adaptation Talks will come to a positive conclusion. Two years of study suggest that there is a great deal of flexibility in the United States' arms control repertoire on the one hand and a loose-enough "fit" between what CFE can deliver and Europe's current security requirements on the other hand, so that it should be possible to preserve the essential elements of CFE as a "security redoubt" against the worst-case turn in Europe's fortunes while adjusting some aspects of the treaty to fit present circumstances more closely. The more likely outcomes from CFE Adaptation will probably feature modest adjustments, perhaps ending the blocs of states and accommodating the security needs of specific signatories. Moderate further TLE reductions might also be in the bargain.

With these possibilities in mind, we counsel against undertaking additional pan-European conventional arms control initiatives. In the near term, useful and practical arms control will probably mean more-local agreements, tailored specifically to address grievances among neighbors. Unless circumstances alter dramatically, Europe-wide negotiations will make little sense, especially in the face of NATO enlargement, in which, presumably, allies will not negotiate arms control pacts with each other.

If the United States were pressed for additional all-Europe arms control, this study indicates that stability and transparency measures could make useful contributions. Deep TLE reductions, coupled with robust confidence- and security-building measures, could also have a major effect, although the expense of the reductions and the current political climate in much of Europe make deep reductions an unlikely choice.

What next? Assuming that CFE modernization is successful, what should the United States consider its next arms control priority? Some unfinished business remains on the European arms control agenda, and there is potential for arms control to help manage the dangers emerging from the Mediterranean basin and Southwest Asia. The unfinished business is to bring the Baltic states into CFE. Membership in the treaty and the CSBM regime would help insulate the Baltic states from pressures by Russia and would provide transparency sufficient to defuse misinformation campaigns, rumors about poor treatment of Russian minorities, and similar issues that might offer Moscow excuses to intervene. As treaty members, these states' annual data declarations would make clear to the world that they have no offensive capabilities and pose no threats to their neighbors. In addition, membership in the two pacts would entitle the Baltic states to information about their neighbors, including prior notice of major exercises and similar activities that, given the small size of these states, could be very dangerous to their prospects of survival as independent, sovereign actors.

The United States should also campaign to bring Slovenia, Finland, Austria, and Sweden under the treaty's aegis. Since their individual circumstances—locations, levels of armaments, histories—are distinct, a thorough examination of the potential issues surrounding their accession to the treaty should be made.

The history of arms control in the Cold War offers some clues about how arms control might be used beyond Europe's borders to improve relations with her neighbors. From the 1958–1959 Geneva Counter Surprise talks through the years of the Mutual and Balanced Force Reduction (MBFR) talks, arms control offered a forum for discussions with the Soviets. In an era when the West was ideologically, economically, and culturally alienated from the Soviet Union, arms control offered venues for talks and discussions. Lenin would have dismissed the whole business as useless "talking shops," but, in practice, arms control provided a sound channel of communication when few others were available to sustain a useful dialogue. Through this channel, the East and West came gradually to know each other better. The United States could pursue the same approach toward its current, more radical opponents in Southwest Asia and on the southern rim of the Mediterranean.

The United States faces similar ideological barriers in and around the Mediterranean basin. Radical Islam, cults of the leader, and similar forces promote barriers to understanding analogous to those encountered with the Soviet Union. It therefore makes sense to use a proven tool, arms control, to try to manage the relationship with these actors and, over time, to reduce their radicalism.

THE NEAR-TERM CONVENTIONAL ARMS CONTROL AGENDA

Even if the establishment of an arms control dialogue like that suggested above proves bureaucratically impracticable, the arms control agenda remains full. For most of Europe and the United States, it could be summarized as follows:

- Continue to participate in the usual arms control processes, consolidating CFE and the CSBM regime by incorporating the Baltic states and the other states.
- That accomplished, move the discussion from arms control to genuine cooperation.
- Build on cooperation fostered in the Partnership for Peace and allied exercises by launching efforts with more teeth.
- Consider mutually beneficial projects, perhaps a regional missile defense system, cooperative anti-submarine-warfare patrolling in the Mediterranean, or accelerated work on NATO's Combined Joint Task Force concept.

The agenda for trouble spots in Europe is as follows:

- The United States should promote solid, local arms control deals.
- As with the Dayton Accords, the United States should use its influence to bring disputing parties together for the negotiations.

NEXT STEPS

The prescriptions for next moves, above, raise immediate questions of *how*: how to incorporate the remaining outlying states, how to manage their individual accession issues, and how to move relations among the states in Europe from an adversarial basis to a truly cooperative one. These ought to be the next steps for the project.